ELECTRICAL ENGINEERING DEVELOPMENTS

AF735976

NETWORKED CONTROL SYSTEMS

RESEARCH CHALLENGES AND ADVANCES FOR APPLICATION

ELECTRICAL ENGINEERING DEVELOPMENTS

Additional books in this series can be found on Nova's website under the Series tab.

Additional e-books in this series can be found on Nova's website under the e-book tab.

ELECTRICAL ENGINEERING DEVELOPMENTS

NETWORKED CONTROL SYSTEMS

RESEARCH CHALLENGES AND ADVANCES FOR APPLICATION

EDUARDO PACIÊNCIA GODOY
EDITOR

nova
science publishers
New York

Copyright © 2018 by Nova Science Publishers, Inc.

All rights reserved. No part of this book may be reproduced, stored in a retrieval system or transmitted in any form or by any means: electronic, electrostatic, magnetic, tape, mechanical photocopying, recording or otherwise without the written permission of the Publisher.

We have partnered with Copyright Clearance Center to make it easy for you to obtain permissions to reuse content from this publication. Simply navigate to this publication's page on Nova's website and locate the "Get Permission" button below the title description. This button is linked directly to the title's permission page on copyright.com. Alternatively, you can visit copyright.com and search by title, ISBN, or ISSN.

For further questions about using the service on copyright.com, please contact:
Copyright Clearance Center
Phone: +1-(978) 750-8400 Fax: +1-(978) 750-4470 E-mail: info@copyright.com.

NOTICE TO THE READER

The Publisher has taken reasonable care in the preparation of this book, but makes no expressed or implied warranty of any kind and assumes no responsibility for any errors or omissions. No liability is assumed for incidental or consequential damages in connection with or arising out of information contained in this book. The Publisher shall not be liable for any special, consequential, or exemplary damages resulting, in whole or in part, from the readers' use of, or reliance upon, this material. Any parts of this book based on government reports are so indicated and copyright is claimed for those parts to the extent applicable to compilations of such works.

Independent verification should be sought for any data, advice or recommendations contained in this book. In addition, no responsibility is assumed by the publisher for any injury and/or damage to persons or property arising from any methods, products, instructions, ideas or otherwise contained in this publication.

This publication is designed to provide accurate and authoritative information with regard to the subject matter covered herein. It is sold with the clear understanding that the Publisher is not engaged in rendering legal or any other professional services. If legal or any other expert assistance is required, the services of a competent person should be sought. FROM A DECLARATION OF PARTICIPANTS JOINTLY ADOPTED BY A COMMITTEE OF THE AMERICAN BAR ASSOCIATION AND A COMMITTEE OF PUBLISHERS.

Additional color graphics may be available in the e-book version of this book.

Library of Congress Cataloging-in-Publication Data

ISBN: 978-1-53613-105-5

Published by Nova Science Publishers, Inc. † New York

CONTENTS

PREFACE

The area of Networked Control Systems (NCSs) has emerged in the last decade as a multidisciplinary research field with the accelerated integration and convergence of communications, computing, and control systems. NCSs are physically distributed and decentralized control systems in which the interconnection among sensors, actuators and controllers is performed using a shared (wired or wireless) communication network. It has become a basic networked control architecture of many nowadays and future applications in Information Technology and automation systems, including the cyber-physical systems, the Industry 4.0, the Internet of Things, and the smart grids and homes.

Even though the NCS offers many advantages over traditional control systems, several challenges also emerged giving rise to many important research focuses. New control strategies to deal with jitter, delays and packet losses, reliability of communications, bandwidth allocation, development of data communication protocols, design of fault detection and fault tolerant systems, real-time scheduling are some of the relative topics extensive studied. Moreover, demands on interoperability, security, real-time performance for networked systems have originated new technological challenges to NCS. Currently, many fundamental questions regarding the control performance and stability of interconnected systems, the effects of shared communication on the performance of control systems, among others remain open.

Advances in wireless sensor networking technology have led to the development of low cost, low power, multifunctional sensor nodes. With these advances, a trend has emerged in using wireless networks in NCS and enabling interoperability between existing wired and wireless systems. These systems are known as Wireless Networked Control Systems (WNCS). However, when wireless networks are used for control applications, such as in WNCSs, a tradeoff between time delays, packet losses and jitter will be required to provide deterministic operation and achieve control and stability requirements. A key performance element for a WNCS is the capability to support real-time applications. Real-time means that the system must be able to response to control requests timely, so that corrections still have their desired effect on the system operation.

A fundamental issue and technological concern in WNCS is the energy consumption of wireless devices. Since the devices are powered by batteries, the lowest possible consumption is required to extend battery lifetime and reduce maintenance during operation. However, reducing energy consumption in WNCS may be challenging due to the today's digital control systems requirement for fast updates. Therefore, research about energy-saving solutions for WNCS applications is gaining importance.

Recently, the Internet of Things (IoT) has attracted significant attention from governments, industry, and academia. The IoT is a networked system that enables the sensing and control of physical systems and a new trend is on the development of IoT-based NCSs. This new type of architecture in which control systems are integrated with IoT infrastructures represents the next evolution of networked control architectures. The concept of cloud computing in industry is growing with virtualized resources distributed as services through the Internet or a standard communication protocol. Even though this idea enables a whole range of novel functionalities, feedback control design and architectures for IoT imposes significant challenges that still needed to be addressed.

This book compiles the last theoretical and experimental results in the topics of NCSs and WNCSs and starts discussing the last trend of IoT-based NCSs. The book is intended for graduate students, researches, and specialists with interest in the study of distributed systems, control over networks and communications problems. The book includes seven chapters written by researches in NCS and addresses some of the challenges and problems discussed. The chapters 1 and 2 are related to the NCS topic. The subject of WNCS is addressed in the chapters 3 to 6 with different application focus from process control to smart grids. The chapter 7 focus on the IoT-based NCS theme.

One important parameter for the design of NCS is the sampling interval of the devices sharing the communication network. It is important to note that different sampling intervals correspond to different amounts of network bandwidth required by the control system. The NCS must sample and transmit data at a sampling interval appropriate to achieve required performance metrics. Moreover, the sampling interval can affect the NCS control performance and stability. The authors of chapter 1 describe the design and evaluation of NCSs with online adaptation of the sampling interval. The idea is centered on a PID controller that online adapts the NCSs sampling interval, reducing the network load while maintaining the NCS performance. Experimental results demonstrate the effectiveness and reliability of the adaptation scheme developed. The authors of chapter 2 propose a compensator design method for NCS, which takes into account delay and sampling intervals variations, guaranteeing stability and a certain performance level under some communication delay and bandwidth constraints. Based on parameter dependent Lyapunov functions, a closed-loop stability condition can be verified and a switching control synthesis is derived from the stability condition. Numerical examples demonstrate the use of the proposed methodology.

The authors of chapter 3 present an overview of WNCS in order to demystify the use of wireless technology on the control loops in the industry. The WirelessHART and ISA100.11a wireless industrial protocols are discussed and compared in terms of control applications. Practical experiments analyzes the behavior of typical WirelessHART network configurations in a level control of a process station. Despite its high degree of applicability, the industrial wireless networks face some technical (reliability, energy consumption, environment interference) challenges for the deployment of large-scale control systems in the industry. In the chapter 4, the authors address the problem of equipment placement, routers as well as sensors and actuators, with focus on control in the industrial wireless networks. A multi-objective mathematical model is proposed to aid the design and deployment of the industrial wireless network.

The authors of chapter 5 present a study of a WNCS in which a Microgrid with a Distributed Generator is controlled remotely via a wireless communication. The methodology is presented by smart grid simulations, where the electrical system was implemented using the software PLECS – blockset, the wireless network was developed using TrueTime – toolbox, both emulated in Matlab®/Simulink®. The chapter evaluated a WLAN and ZigBee network to control the generation of the active power under different network parameters, such as the loss probability and distance between devices. In chapter 6, the authors address a critical factor related to implementation of WNCSs that is the energy source of the equipment, which have limited lifetime dependent on the amount of access and data transmitted. The chapter presents the study and development of a multi-rate model predictive control (MPC) as an alternative to improve energy efficiency in industrial applications of WNCSs. The proposed strategy uses less process variables data as it does not requires frequently updates of the process variables transmitted by the wireless instruments.

Finally, chapter 7 provides an overview about the integration between industrial automation and information technology demanded by the new applications of the Industry 4.0. In this context, the chapter presents the proposal of an IoT-based networked control architecture based on integration of a WNCS to a Web server in which the control devices communicate via COAP protocol. A case study using the Hardware-In-the-Loop (HIL) technique evaluates the closed loop control of a level plant through a modified PID controller with communication via Web Service RESTful with the Web server.

I hope you enjoy the book and its content may contribute with your research and development. Thank you for your interest.

Eduardo Paciência Godoy, PhD

In: Networked Control Systems
Editor: Eduardo Paciência Godoy
ISBN: 978-1-53613-105-5
© 2018 Nova Science Publishers, Inc.

Chapter 1

NETWORKED CONTROL SYSTEMS WITH SAMPLING INTERVAL ADAPTATION

Eduardo Paciência Godoy[1,*]
and Arthur José Vieira Porto
[1]São Paulo State University (Unesp), Sorocaba, SP, Brazil
[2]University of São Paulo (Usp), São Carlos, SP, Brazil

ABSTRACT

Networked control systems (NCSs) are distributed control system where the sensors, actuators and controllers are physically separated and connected through communication networks. The challenge in the development of NCS is to overcome the effects of the network delays, packet losses and message sampling intervals on the NCS performance and stability. This chapter describes the design and evaluation of NCSs with online adaptation of the sampling interval. The idea is centered on a PID controller that online adapts the NCSs sampling interval, reducing the network load while maintaining the NCS performance. The performance and robustness of the online adaptation scheme are analyzed for different NCS configurations used on a CAN-based NCS research platform,

* Corresponding Author Email: epgodoy@sorocaba.unesp.br.

showing significant results in the CAN network load reduction. Experimental results led to the conclusion that the developed online adaptation scheme is reliable for application to NCS, providing an acceptable performance for the NCS even in the worst-case scenario investigated.

Keywords: CAN protocol, network load, PID controller, sampling interval

INTRODUCTION

Networked Control Systems (NCS) have become a widely research topic due to the challenges and opportunities of developing a system merging knowledge of three fields: control theory, communication networks and real-time systems (Gupta & Chow, 2010). NCS represents the evolution of networked architectures for communication and control (Sauter, 2010) and this technology differs from the traditional fieldbus systems in that the controller and the plant, sensors and actuators are physically separated and connected through an industrial network (Gupta & Chow, 2010).

Even though the NCS offers many advantages over traditional centralized control systems, several challenges also emerged, giving rise to many important researches (Heemels et al., 2010). New control strategies to deal with jitter, delays and packet losses, reliability and security of communications, bandwidth allocation, development of data communication protocols, design of fault detection and fault tolerant systems, real-time information collection and scheduling are some of the relative topics extensive studied (Baillieul & Antsaklis, 2007).

Another important parameter that affects the NCS performance and stability is the message sampling interval (Moyne & Tilbury, 2007). The NCS must sample and transmit data into messages at a sampling interval appropriate to achieve required performance metrics. However, if this sampling interval is higher than the network bandwidth available, the network becomes overloaded, originating additional network delays and jitter, and causing packet losses and errors in the transmissions of messages (Lian, Moyne & Tilbury, 2002).

The development of a new control strategy able to cancel, or at least minimize these effects and ensure the required stability and control requirements (Martins & Jota, 2010) is mandatory. This is necessary so that all the advantages of the NCS may be used. An important stage in the development of control strategies for NCSs, which is sometimes neglected, is its robustness assessment (Penna, Intrigila & Magazzeni, 2009). The robustness is the ability of a control strategy to maintain good and acceptable performance for the NCS even in the presence of significant disturbances or NCS parameter variations. This type of analysis is essential for NCS due to the impact on its performance and stability of several different parameters such as network delays, packet losses and message sampling interval.

Following this guideline, this paper presents the development and application of NCSs with online adaptation of the sampling interval. The online adaptation scheme is based on a discrete time PID controller that automatically selects the NCS sampling interval in accordance with the NCS output. The main benefit of this strategy is the significant decrease of the network load, hence diminishing the network delays and packet losses caused by the waiting time for message contention. The robustness of the sampling interval adaptation scheme is evaluated through the performance analysis for different NCS configurations and conditions of the CAN network load (the used percentage of the total capacity to transmit messages on the network). Experimental results on a CAN-based NCS research platform led to the conclusion that the developed sampling interval adaptation scheme is reliable for application to NCSs, ensuring an acceptable performance for the NCSs even in the worst-case scenario analyzed. An important contribution of this paper is the experimental part, not only because most of the published works only focused on the theory development for NCS, but also because it is meaningful for the real implementation and practical application of NCS.

This chapter is organized as follows. After this introduction, section II presents a literature review about the development of control strategies for NCS, discussing related works and exposing the state of the art in this topic. An explanation about the sampling interval problem in NCS and a detailed description of the online adaptation scheme are presented in Section III. In

Section IV, some aspects of the developed and used experimental NCS platform are discussed. The sequence and description of the tests are presented in Section V in which experimental results for performance evaluation are given and discussed. Conclusions are outlined in Section VI.

CONTROL STRATEGIES FOR NCSs

NCSs present new challenges for the application of traditional design and control techniques (Baillieul & Antsaklis, 2007). For NCSs, the control design shall consider, simultaneously, several inherent factors to NCS such as the effects of sampling interval, time variant network delays and loss of information on the network (Heemels et al., 2010). For most cases, new control strategies are required for NCSs to mitigate the effects of these factors while maintaining its performance and guaranteeing its stability. Tipsuwan & Chow (2003), Hespanha, & Naghshtabrizi, Xu (2007) and Vatanski et al., (2009) are well-known revision papers about NCS control strategies. A problem related to some control strategies is that they present, at most times, complex solutions with algorithms that require high processing and use information about the network, which are sometimes difficult to obtain.

The most common approach for the design and implementation of NCS consists in the periodic execution of a control algorithm, by defining a fixed execution period (time-driven) for the controller or by defining a fixed sampling interval for the sensor messages (event-driven) (Gupta & Chow, 2010). Both techniques use in a static manner the available bandwidth for data communication without considering other factors such as the momentary network load and variations on the controlled systems. However, results of recent papers have proved that using dynamic definition techniques of this execution period may result in a better control performance for the NCS when compared to using consolidated techniques for static definition or fixed execution period (Cervin et al., 2010; Heijmans et al., 2017). Two current trends can be identified for control strategies that use dynamic definition of this execution period (Camacho et al., 2010).

The first tendency uses techniques to change the controller execution period in accordance with the dynamics of data measured from the NCS such as the network delays, packet losses and the network load. The main goal of these approaches is to improve the NCS control performance using efficiently all available network bandwidth for data transmission. Martí et al., (2010) present a technique that can improve the performance of a set of CAN-based NCS by additionally transmitting non-periodic control messages, proportionally to the CAN network bandwidth available in that moment. The application of a control strategy to maximize the CAN network load to improve the system performance is also presented in Godoy et al., (2010).

The second tendency uses event-based techniques for the controller triggering, which originates non-periodic execution of the controller and therefore non-periodic message transmissions. The goal of these methods is to decrease the network load at the same time ensuring control performance and stability requirements (Araujo et al., 2014). Camacho et al., (2010) present an experimental investigation about the development and implementation of self-triggered controllers for NCS. In this paper, in addition to standard tasks performed every execution cycle, the controller also uses an algorithm to calculate when the next execution cycle will be triggered. The results of the paper show similar control performance for the NCS with self-triggered controller in comparison to a periodic controller, yet obtaining a small decrease in the CAN network load.

In order to handle the effect of the message sampling interval in NCS and using the ideas of the cited techniques for dynamic definition of the controller execution, in this paper an online adaptation scheme was developed. Adapting or adjusting the NCS sampling interval has been a successful control approach for reducing the impact of network problems (Morawski & Ignaciuk, 2016). The adaptation strategy shares the same goal of the event triggering technique that focuses on decreasing the network load. Nevertheless, it innovates and differs from the cited since it uses the concepts of another technique (Cervin et al., 2010), which dynamically defines the sampling interval in accordance with the NCS output. This concept is used to achieve a greater reduction in the network load while maintaining the NCS performance, which represents an advantage of the online adaptation scheme, developed in this paper against the others.

The online adaptation scheme is centered on a discrete time PID controller that automatically adapts the NCS sampling interval according to its output data. With the dynamic definition of the message sampling interval transmitted by the sensor, a dynamic change on the controller execution is obtained at the same way. As a result, a dynamic change in the transmission of control messages to the NCS actuator is also obtained. This dynamic definition in the transmission of messages in the NCS therefore provides a reduction in the network load and minimizes possible degrading effects while maintaining the required control performance. In this paper, this adaptation scheme developed for NCS is also evaluated to figure out its reliability, efficiency and robustness of application even under worst-case condition of operation.

Sampling Interval Adaptation Scheme for NCS

Sampling Interval x Performance

Lian, Moyne & Tilbury (2002) define the relationship between the sampling interval and the performance of a NCS as one of the main concepts to study and analyze. As presented in Moyne & Tilbury (2007), the Figure 1 shows a diagram to assist the selection of this design parameter and to visualize overall NCS performance at different sampling intervals. For the understanding of this diagram, Moyne & Tilbury (2007) explain that the worst, unacceptable and acceptable regions as well as the points A,B and C may be defined according to required control system specifications such as overshoot, response time, tracking and steady state error and phase margin. The performance of a continuous control system is independent of the sample time. The performance of a digital control system approaches the performance of a continuous time system as the sampling interval becomes faster. For the networked control case, the performance is worse than the digital control system at low frequencies (point A), due to the network delays in the NCS. Smaller sampling intervals may be needed in order to guarantee a certain level of control performance but the network will present high level of idleness (point B). As the sampling interval gets faster, the network load becomes heavier and the idleness reduces, the possibility of more contention time and data loss increases in a bandwidth-limited network, and longer delays result. Point C is the situation when the network is becoming saturated. Between points B and C lies the best operation range where the sampling interval is optimized to the control and networking environment. Due to the interaction of the network and control requirements, the selection of the best sampling interval is a compromise (Moyne & Tilbury, 2007).

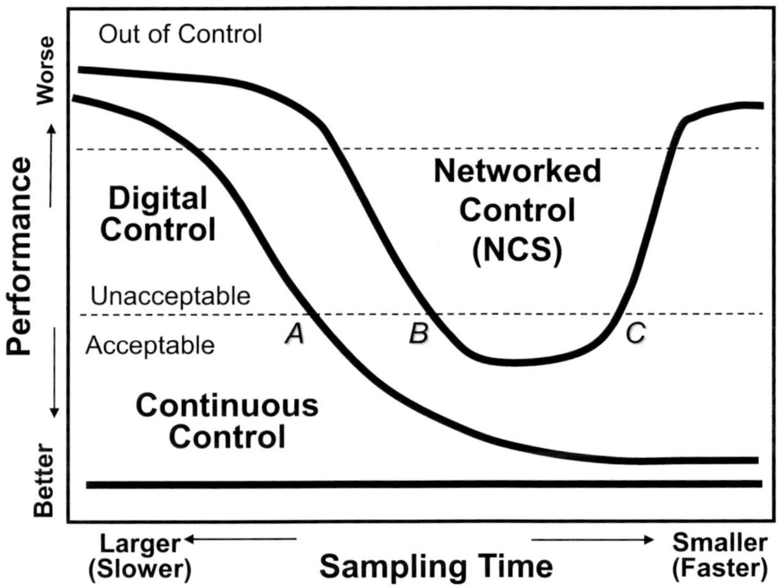

Figure 1. Relationship between Sampling interval and Performance for Different Types of Control (Moyne & Tilbury, 2007).

For NCS using low bandwidth networks such as CAN (when compared to the Industrial Ethernet gigabit bandwidth), the message sampling interval problem is even more complicated. Moreover, it is not only because the great impact of this parameter on the performance of the NCS, as also by this parameter being quite related to the CAN network load. If the NCS sampling interval is very fast, there may be saturation of the communication network. In this situation, where the CAN network presents high load and becomes overloaded, large network delays are induced and errors in transmitting messages become constant, degrading the performance and may make the NCS unstable.

Therefore, it is important to deal correctly with the sampling interval when developing NCSs. In order to accomplish this goal, in this paper was developed an online adaptation scheme, based on a PID controller, which is able to adapt the NCS sampling interval in accordance with the NCS output and a few user defined parameters. The main desired objective of the adaptation idea is to reduce the network load while maintaining the NCS performance.

Discrete Time PID Controller as a Basis

According to Eriksson (2008), controllers for NCS cannot be designed with continuous time control theory because the resulting performance is unsatisfactory. The controllers for NCSs have to handle the network delay effects in the systems and need to be designed in discrete time. The controller used as a basis for the adaptation scheme is a discrete-time PID derived with the backward derivative approximation with setpoint weighting, reference off, filtering on derivative part and Anti-Windup of the integrative part (Åström & Hägglund, 1995). Figure 2 presents the block diagram of the defined controller with sampling interval (*h*), filtering constant of the derivative part (*N*), weighting constants (*B* and *C*) and the Anti-Windup parameter *Tt* equal to $\sqrt{T_i.T_d}$ for PID controllers and equal to *Ti* for PI controllers.

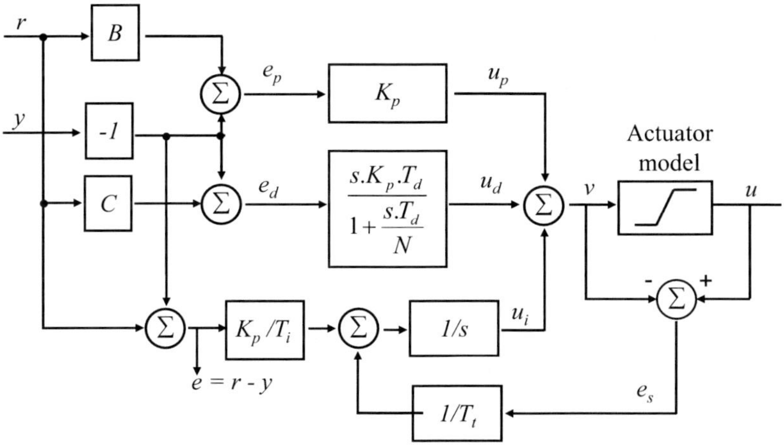

Figure 2. Block Diagram of the Defined PID Controller for NCS.

In the setpoint weighting technique (Åström & Hägglund, 1995), the system overshoot for setpoint changes is less for smaller values of parameter *B* that adjusts the proportional part. The parameter *C* is usually zero, a technique called reference off, which uses only to process output (y) as input to the derivative part, to avoid major changes in the control signal due to sudden changes of setpoint. The parameter *es* is the difference between the nominal controller output (*v*) and the saturated controller output (*u*) according to the actuator model. If *es* is zero, there is no saturation and it will not have any effect on the integrator operation. When the actuator saturates, the signal es is different from zero and it will drive the integrator output to a value such that the signal *v* is close to the saturation limit.

Detailing the Adaptation Scheme (AS)

The principle of operation and the algorithm of implementation of the adaptation scheme are shown in Figure 3. It is important to understand that the sampling interval adaptation scheme starts automatically with a change at the NCS setpoint or if the NCS output is outside the operation range.

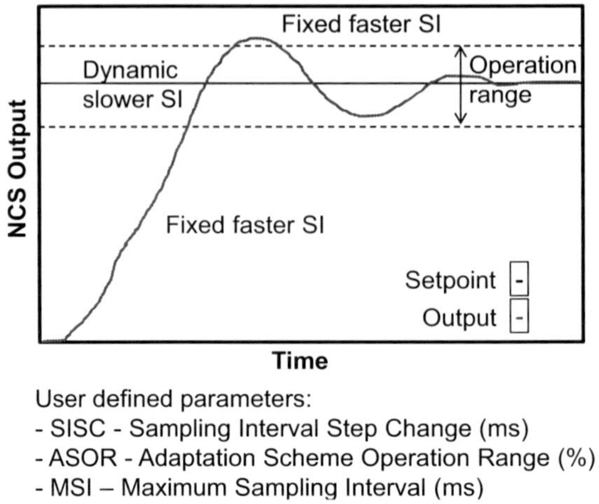

Figure 3. Principle of Operation of the Adaptation Scheme (AS) of the Sampling Interval for NCS.

NCS_SI = IS; % Current NCS sampling interval is equal to the initial faster sampling interval
While (NCS_in_Operation)
{
Get (process variable); % receive sensor message
If (NCS_Output =< ASOR)
{
If (NCS_SI < MSI)
NCS_SI = NCS_SI + SISC;
Else
NCS_SI = IS;
}
Update (NCS_SI); % transmit the new value of the NCS sampling interval in a message to the sensor
}

Figure 4. Algorithm of the Adaptation Scheme (AS) of the Sampling Interval for NCS.

According to the Figure 3, the user needs to set three parameters for the adaptation scheme. The maximum sampling interval (MSI) defines the maximum SI that can be used in the NCS. The sampling interval step change (SISC) in milliseconds (ms) defines how fast the changes in the SI will be made at each controller period. The adaptation scheme operation range (ASOR) in percentage (%) determines the area related to the NCS setpoint in which the AS will operate. There is no relationship between the ASOR (%) parameter and the defined allowable error in the steady state response of the control system.

As can be seen in the algorithm of Figure 4, during the NCS operation, the initial SI value (NCS_SI) is faster in order to achieve an initial good transient output and performance. If the NCS output (NCS_Output) is within the area designated by the operation range (ASOR), the AS automatically reduces the NCS sampling interval (increasing the SI value and making it slower - NCS_SI = NCS_SI + SISC) while maintaining the performance of the NCS. Consequently, the NCS network load will decrease too. If the NCS output is outside the designated area (operation range), the AS performs the contrary action, reinitializing and increasing the NCS sampling interval to the initial value (decreasing the SI value and making it faster - NCS_SI = IS). The SI change implemented by the AS is performed using messages transmitted in the network (Update NCS_SI). The NCS controller computes the new SI values and these new values are transmitted to the sensor using the control messages (additional two bytes for the new SI value). As soon as the sensor receives a new SI value, it resets the current SI (that it is using) and starts sending the process variable message at each SI defined by the received value.

Experimental Setup: Networked Control System (NCS) Platform

The platform used for the experiments in this paper was developed for NCS research and experimentation (Godoy et al., 2010). The communication network used for the control systems integration and information exchange is the CAN protocol. The architecture of the NCS platform is shown in Figure 5. Common control systems used in the industrial area such as DC motor velocity and position control, temperature control, tank level control and belt conveyor control are selected for the platform.

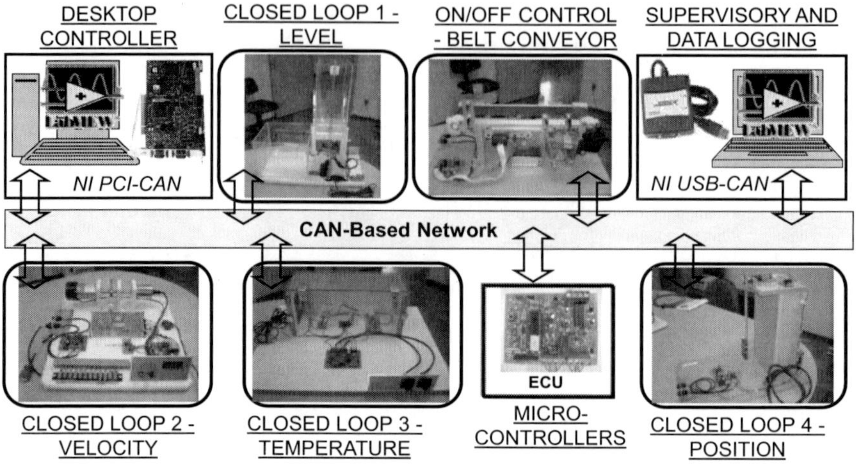

Figure 5. Experimental Setup: CAN-Based NCS Research Platform.

Each of the defined systems has a microcontroller based electronic control unit (ECU) which is responsible for the data acquisition (sensor), actuation in the plant (actuator) and communication with the CAN-based network. A desktop with LabVIEW and PCI-CAN interfaces from National Instruments are used for the development of the control strategies of the NCS. The architecture developed has high flexibility for the research and teaching of NCS, providing capabilities to perform several NCS tasks such as analysis, modeling, simulation and control.

For each of the NCS that composes the platform, the time-driven sensor node samples the plant or process periodically and sends the samples to the controller node (desktop computer) over the CAN network. Upon receiving a sample, the controller computes a control signal that is sent to the actuator node, where it is subsequently actuated. The threads executing in the controller and actuator nodes are both event-driven. All the closed loop control systems in the platform are sharing both limited CAN network bandwidth and controller CPU. The competition for these constrained resources will certainly increase the network delays of the control loops and degrade the overall performance of the NCS in the platform.

RESULTS AND DISCUSSIONS

Adaptation Scheme Performance Evaluation

In order to evaluate the versatility and robustness of the adaptation scheme (AS) and demonstrate its benefits of application, several experiments were conducted for different configurations of NCS using the NCS platform. For all tests performed, the five NCS of the platform were kept in operation, in order to achieve a situation with significant message traffic (periodic and non-periodic) on the CAN network, consistent with the reality of networked industrial applications. In addition, this configuration assures a bigger utilization of the CAN network and enhances the results analysis. An initial message sampling interval (SI) of 10ms was configured for the sensors of the NCS, while the belt conveyor NCS has an event based control (presence of pieces).

Table 1. Parameters used in the evaluation experiments of the adaptation scheme

NCS	Velocity	Position	Level	Temperature
Parameters of the Adaptation scheme				
Maximum SI (s)	0,2	0,2	0,4	0,8
SI Step Change (ms)	5	5	5	10
AS Operation Range (%)	10	5	5	10
Parameters of the Discrete Time PID Controllers				
Proportional Gain - K	0.07	3	10	15
Integrative Time – Ti	0,65	-	0,55	7,5
Derivative Time - Td	-	0,001	0	0,1

The parameters defined for the AS and for the PID controller used in each NCS of the platform are presented in Table 1. Additionally, the following parameters were used: CAN network speed of 250 kbit/s, message priorities sequence (position = 1, velocity = 2, temperature = 3 and level = 4), messages data length of two bytes for the sensor and four bytes (two for the control signal and two for the new sampling interval) for the controller, a value of N = 10 for the constant filtering, C = 0 and B = 1 for the weighting constants.

Figure 6 explains the results obtained with only the temperature control NCS using the adaptation scheme (AS) for the sampling interval (SI). Figure 6 presents, in the top graph, the output performance for the NCS during the operation of the AS, and at the bottom one, the variations in the NCS SI and in the CAN network load.

As can be seen at the bottom graph, the SI of the NCS for temperature control was automatically increased until 800ms since the AS started operating. With this increase in the NCS SI, the CAN network load decreased from approximately 20% to 11%, while the NCS performance was kept constant, as shown in the NCS step response graph at the top graph in Figure 6. These graphs provides a good information about the operation of the developed AS.

It is important to verify that on the instant t = 38s in Figure 6, as the NCS output is not within the AS operation range, the NCS SI is reinitialized to its initial value (10 ms), providing the great peak on the CAN network load graph (green) and the initialization on the NCS sampling interval graph (blue). Another finding is that the increase on NCS SI was not linear or the curve was not like an integrator. It can be explained by the nonlinearities on the network such as the delays, as the adaptation scheme uses the network to transmit messages to change the sampling interval.

With the application of the AS simultaneously for more than one NCS, the graph of the variation on the CAN network load will show the changes during the experiment time. So, each listed change will be related to the instant (t) in which the AS started operating on a given NCS, automatically managing its sampling interval (SI). The first experiment was conducted with the application of the AS, at the same time, for all the four NCS of the platform. The compiled results presented jointly in Figure 7 demonstrate the efficiency of the AS by a significant reduction in the CAN network load.

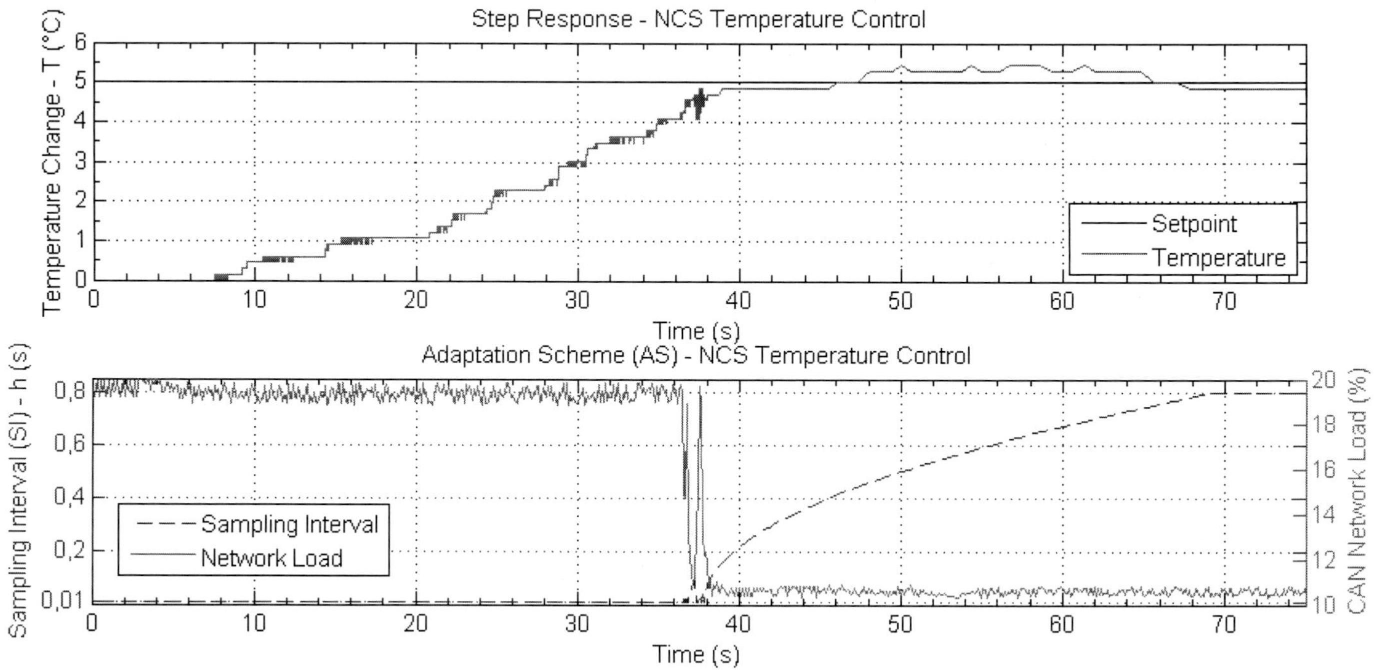

Figure 6. Adaptation Scheme (AS) for CAN-Based NCS: NCS Step Response, Variation in the Sampling Interval and Variation in the CAN Network Load during the NCS Operation.

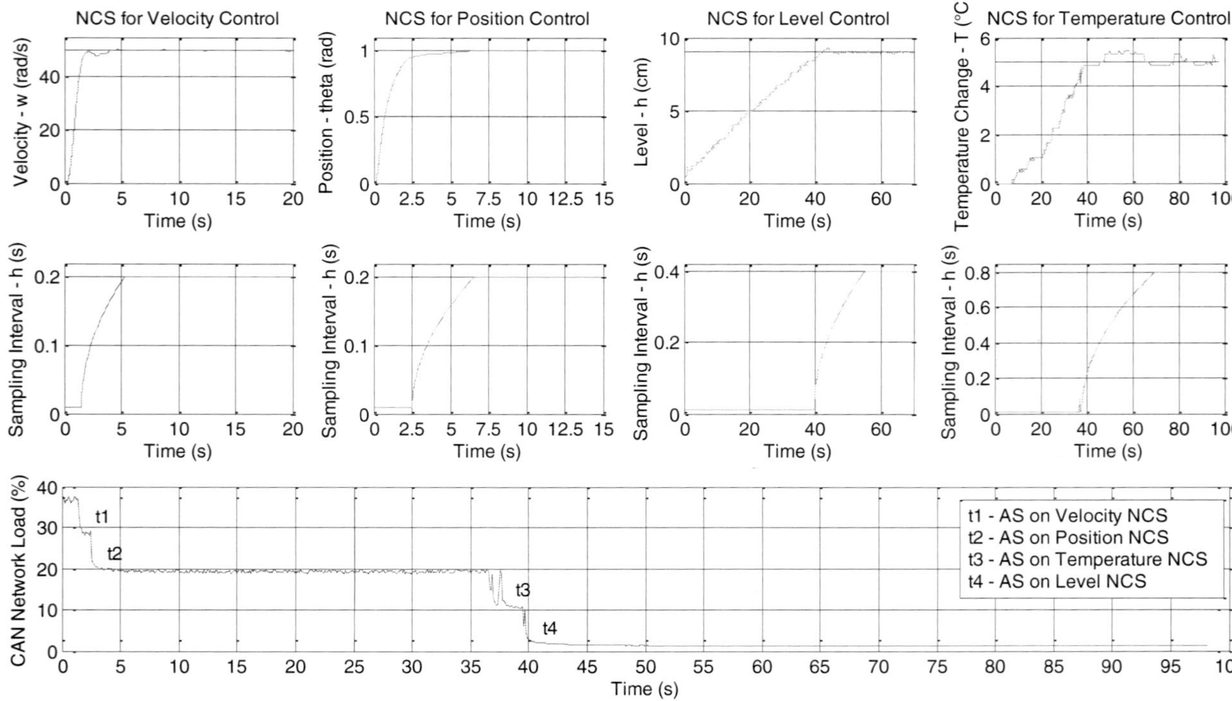

Figure 7. Application of the Adaptation Scheme (AS) for all the NCS of the Platform: NCS Step Responses, Sampling Interval (SI) Variations and CAN Network Load Variation.

Figure 7 presents for the four NCS of the platform, a graph showing the NCS control performance to a step response (top graphs) and a graph showing the change of the NCS sampling interval (intermediate graphs) during the operation of the AS. In addition, a graph showing the variation in the CAN network load during the whole experiment (bottom graph) is also presented.

In accordance with the Figure 7 four different changes (t1, t2, t3, t4) can be seen at the CAN network load graph, which are related to the instants in which the AS began to operate for each of the NCS (t1 for velocity, t2 for position, t3 for temperature and t4 for level). Summing the reduction effect on the network load obtained for all NCS when using the AS, the result is even more significant. In this case, the reduction in the CAN network load was from 37% to 2%, without sacrificing any NCS control performance or reliability. The results obtained with this experiment prove the reliability and the great versatility of the AS application for different types of NCS.

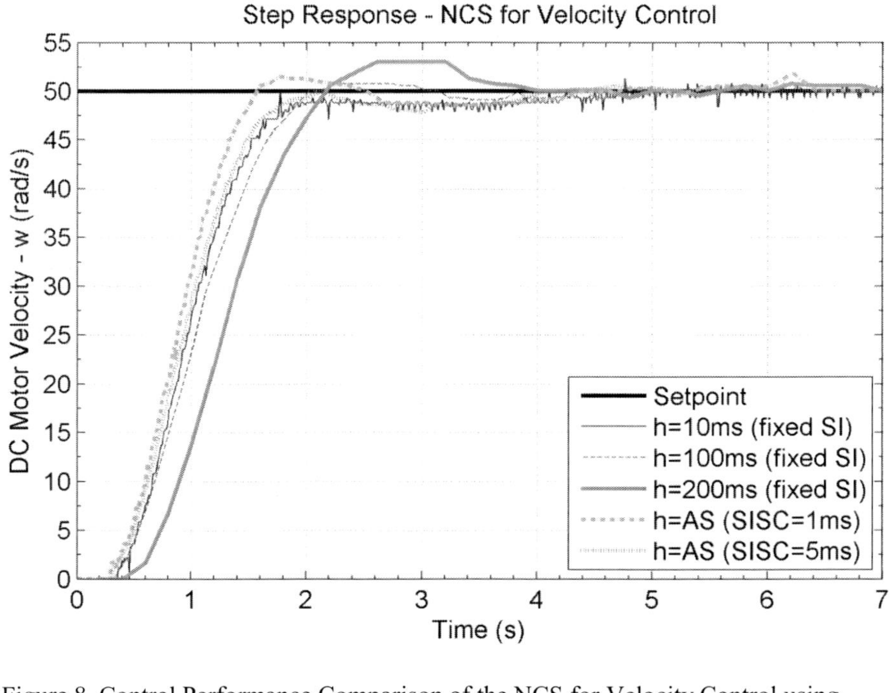

Figure 8. Control Performance Comparison of the NCS for Velocity Control using fixed Sampling Intervals and with the Adaptation Scheme (AS).

Figure 8 presents the comparison among the control performance for different sampling intervals including fixed values and the adaptation scheme (AS) developed for velocity control NCS. The objective of this analysis is to check the difference among the fixed and variable approaches for the NCS sampling interval. In addition, to use the results to verify if there is a disadvantage in the application of the AS implemented.

It is possible to verify through the curves of Figure 8 that there is no disadvantage or performance degradation in using the adaptation scheme (AS) instead of the fixed sampling intervals. On the contrary, a slight improvement in the NCS time response (rise time) can be noticed, mainly in the case of the AS with SISC (sampling interval step change – AS parameter) of 1ms. The comparison results for the others NCS provided the same conclusion and were omitted for simplification. The obtained results prove the efficiency of the AS for application in NCS.

Adaptation Scheme Robustness Evaluation

The application of the adaptation scheme (AS) is important by the fact that usually in NCS, the network characteristics, such as network delays, are time variant (Lian, Moyne & Tilbury, 2002). Because of that, a NCS operating with high (or very slow) fixed sampling intervals can at some point lose its control requirements and become more oscillatory (Heemels et al., 2010). On the other hand, if the NCS sampling interval is too small (or fast), the CAN network presents high load and becomes overloaded, inducing larger network delays (waiting time for message contention).

To demonstrate the importance of the AS for NCS and to evaluate its robustness of application, an experiment was carried out comparing the results for the NCS of the platform to four different configurations and CAN network conditions. Table 2 resumes the parameters and conditions of operation for the experiments. The experiment 3 is the same experiment whose the results are presented in Figure 6.

Table 2. Experiments for robustness evaluation

Experiment	NCS Sampling Interval Approach	CAN Network Condition	CAN Network Load
1	Fixed h = 10ms	Without extra traffic	37%
2	Fixed h = 10ms	With extra traffic = 55% of the CAN load	92%
3	Using AS	Without extra traffic	37% to 2%
4	Using AS	With extra traffic = 55% of the CAN load	92% to 57%

According to Table 2, for each experiment conducted two main configurations were used. The NCS sampling interval (SI) defines the approach used for the selection of the SI for the NCS of the platform. Moreover, the CAN network condition that defines the insertion of an extra message traffic on the CAN network equal to approximately 55% of the network capacity. This extra traffic was obtained with the connection in the CAN network of a computer configured to transmit repeatedly extra messages increasing the CAN network load. The objective of this extra traffic is to evaluate the NCS operating with the AS in severe conditions of the CAN network.

The experiment 1 was performed with a fixed SI of h=10ms for all NCS and without extra traffic on the CAN network. In this situation, the CAN network load measured was equal to 37% during the experiment. In the experiment 2, the same fixed SI of h = 10ms was used but with the extra traffic in the CAN network. In this situation, the CAN network load during the experiment was equal to approximately 92%, configuring a severe condition of operation for the NCS of the platform. The experiments 3 and 4 have repeated the experiments 1 and 2 with the application of the AS. The idea is to compare the results to evaluate the benefits of the AS application. It is important to show that the experiment 4 represents the worst-case scenario for application of the AS. In this experiment, the CAN network load is initially 92% and all the four NCS of the platform use the AS simultaneously, originating a great change in the transmission of messages.

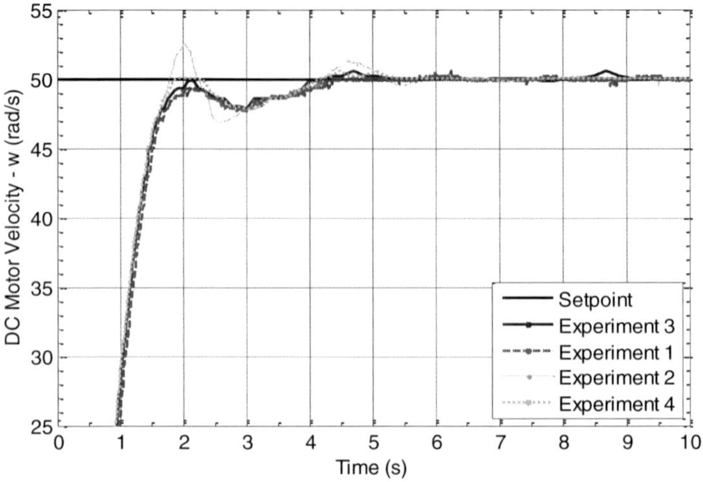

Figure 9. Performance Comparison of the NCS for Velocity Control for the Experiments in Table 2.

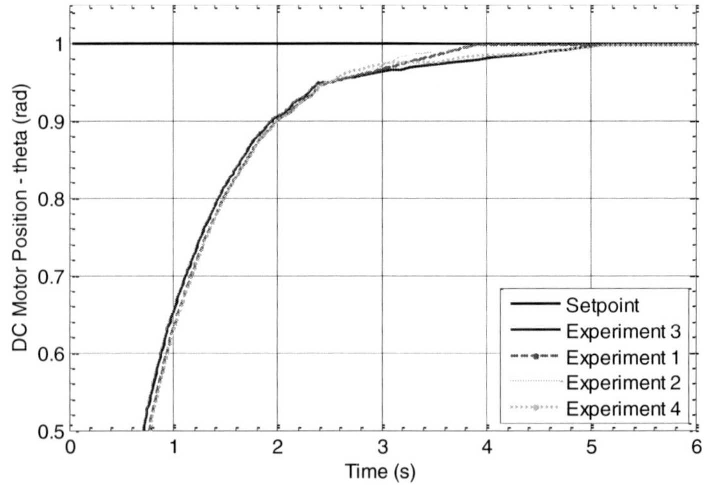

Figure 10. Performance Comparison of the NCS for Position Control for the Experiments in Table 2.

The experiment 2 (h = 10ms/with extra traffic), carried out with the NCS operating with a severe condition of 92% of the CAN network load, demonstrated that the degradation in the NCS control performance is more pronounced according to the occupation of the CAN network. This can be explained by the great network delays, induced mainly by the messages contention and blocking time to access the CAN network. Observing the graphs of the Figure 9 and Figure 11, the results of experiment 2 show that there is deterioration in the control performance of the NCS for velocity and temperature control, which had not been affected in the previous experiment 1. In the same way as in the experiment 1, the NCS for level control became more oscillatory, tending to instability.

The results of the experiments 3 (h = AS/without extra traffic) and 4 (h = AS/with extra traffic) showed the importance of the adaptation scheme (AS) developed for NCS. The application of the AS for the NCS provided an automatic way to overcome the influence of the sampling interval (SI) in the performance of the NCS using the dynamic definition of this parameter in accordance with the NCS output. In addition to this change on the NCS sampling intervals and consequently with the reduction in the CAN network load and network delays, the AS has also provided a way to guarantee the control performances for the NCS.

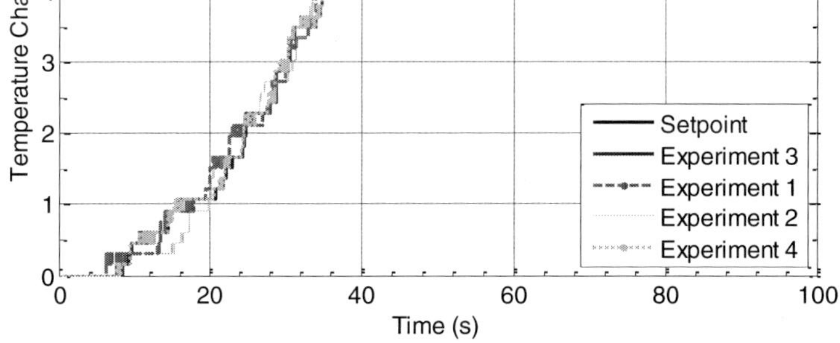

Figure 11. Performance Comparison of the NCS for Temperature Control for to the Experiments in Table 2.

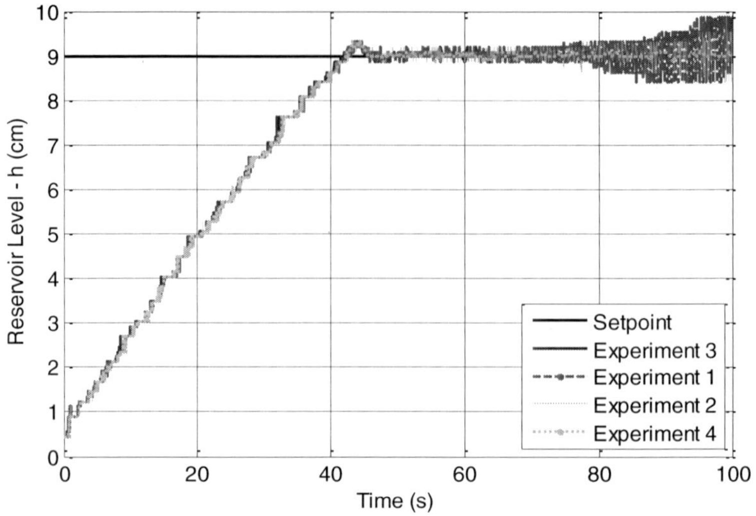

Figure 12. Control Performance Comparison of the NCS for Level Control for the Experiments in Table 2.

It is important to show that the experiment 4 (h = AS/with extra traffic) represents the worst-case scenario for the AS application, in which all the NCS operates simultaneously with the AS and the CAN network load is almost its maximum capacity. Observing the graphs of the Figure 9 to 12 related to the experiment 4, the application of AS provided an acceptable control performance for all the NCS even with the initial occupation of 92% of the CAN network load. Comparing these results with the results of the experiment 2 in the same figures, the application of the AS improved the control performances of the NCS, which had been degraded by the use of fixed sampling interval in experiment 2. These results demonstrate the robustness of the AS for NCS even for the worst-case condition of the CAN network used. The automatic selection of the NCS sampling intervals during the experiment 4 originated a reduction of the CAN network load from 92% to 57% (representing the 55% related to the extra message traffic plus 2% related to the control traffic of the NCS). For the experiment 3, which the compiled results are presented in the Figure 6, the reduction on the CAN network load was from 37% to 2%.

FINAL REMARKS

In a nutshell, it could be verified by the experimental results that the use of fixed sampling intervals (SI) with the increase of the CAN network load degrade the control performance of NCS, making it more oscillatory. This conclusion can be seen in the graphs of the Figures 9, 11 and 12. However, the results of the Figure 10 showed that the NCS for position control has not presented significant performance degradation related to the increase in the CAN network load. It is explained because the CAN messages of this NCS have the highest priorities to access the CAN network of the platform, which consequently means smaller network delays. This fact may lead to another conclusion that the prioritization of messages available in some industrial networks such as CAN may also affect the performance of NCS.

The verified sampling interval problem in NCS though could be overcome with the application of the adaptation scheme (AS) developed in this paper. The results obtained from the experiments prove the reliability and the benefits of the application of the AS. In addition, the results evidence the versatility and robustness of the AS for application in NCS. With the use of the AS, a significant reduction (maximum of 38%) in the CAN network load was obtained during the operation of the NCS in the platform, minimizing potential problems related to network delays and information lost on the network which could affect the performance and stability of the NCS.

Another advantage of the AS is its full compatibility with industrial systems already deployed. New NCS can be developed using the AS on industrial systems, which use NCS with fixed sampling intervals, or even on distributed systems with industrial networks that do not have solutions based on the NCS technology.

Future work about the adaptation scheme for the sampling interval in NCSs aims to further investigate the impact of different user defined parameters on the NCS performance and study the AS stability analysis.

CONCLUSION

A new adaptation scheme (AS) was developed in order to deal with one of the principal problems related to NCS that is the selection of the message sampling intervals. Sampling intervals (SI) not selected correctly along with network delays and packet loss can difficult the controller design and degrade the NCS performance and stability. To overcome this problem, an online adaptation scheme was developed based on an adaptive discrete-time PID controller that has the ability to change the message SI (and consequently the controller parameters) in accordance with the NCS operation, reducing the network load and maintaining the NCS performance. The main benefit of the developed AS is the automatic management of the message SI and significant decrease in the network load, hence diminishing the network delays and packet loss caused by the waiting time for message contention and access to the network, which can affect the NCS performance.

The developed AS was implemented, tested and evaluated through a series of experiments conducted in a real CAN-based NCS research platform. The AS versatility was verified through the performance analysis of different types of NCS using the AS. The AS can be applied to more than one NCS simultaneously and is fully compatible with industrial systems that do not have solutions based on the NCS technology. The results also demonstrated that the performance and stability of NCS using low bandwidth networks such as CAN might be degraded by the increase in the network load, which induces greater network delays and causes packets losses.

The robustness of the AS was also verified against different NCS configurations and network loads. The experiments were conducted under extreme conditions of operation including the worst-case scenario in which all NCS are using the AS at the same time with the maximum CAN network load (92%). In all scenarios investigated, the AS was efficient to guarantee the NCS performance and stability when compared to the results of the NCS using fixed sampling intervals.

The results presented led to the conclusion that the developed adaptation scheme (AS) is reliable for application to NCS, providing a solution to overcome two verified problems in NCS that are the message sampling interval and network load impact on the performance and stability of this system.

ACKNOWLEDGMENT

Research supported by grant 2011/23217-0, São Paulo Research Foundation (FAPESP).

REFERENCES

Åström, K. J., Hägglund, T. (1995). *PID Controllers: Theory, Design, and Tuning*, 2nd ed., Instrument Society of America.

Araújo, J., Teixeira, A., Henriksson, E., Johansson, K. H. (2014). A down-sampled controller to reduce network usage with guaranteed closed-loop performance. In *53rd IEEE Conference on Decision and Control*, Los Angeles, CA, 2014, pp. 6849-6856.

Baillieul, J.; Antsaklis, P. J. (2007). Control and Communication Challenges in Networked Real-Time Systems. *IEEE Proceedings of the Technology of Networked Control Systems*, 95, 1, 9-28.

Camacho, A.; Martí, P.; Velasco, M.; Lozoya, C.; Villà, R.; Fuertes, J. M.; Griful, E. (2010). Self-triggered networked control systems: An experimental case study. In *IEEE International Conference on Industrial Technology* (ICIT), pp. 123-128.

Cervin, A.; Velasco, M.; Martí, P.; Camacho, A. (2010). Optimal Online Sampling interval Assignment: Theory and Experiments. *IEEE Transactions on Control Systems Technology,* 99, 1-9.

Eriksson, L. (2008). *PID Controller Design and Tuning in Networked Control Systems*, PhD Thesis, Department of Automation and Systems Technology, Helsinki University of Technology.

Godoy, E. P.; Pereira, R. R. D.; Scorzoni, F.; Porto, A. J. V.; Inamasu, R. Y. (2010). CAN-based Platform for the Study and Experimentation on Networked Control Systems (NCS). In *IFAC Symposium on Mechatronic Systems* (IFAC MECH).

Godoy, E. P.; Sousa, R. V.; Porto, A. J. V.; Inamasu, R. Y. (2010). Design of CAN-Based Distributed Control Systems with Optimized Configuration. *Journal of the Brazilian Society of Mechanical Sciences and Engineering,* 32, 420-426.

Gupta, R. A.; Chow, M. Y. (2010). Networked Control System: Overview and Research Trends. *IEEE Transactions on Industrial Electronics*, 57, 7, 2527-2535.

Heemels, W. P. M. H., Teel, A. R., Van De Wouw, N., Dragan, N. (2010). Networked Control Systems with Communication Constraints: Tradeoffs between Transmission Intervals, Delays and Performance, *IEEE Transactions on Automatic Control*, 55, 8, 1781-1796.

Heijmans, S. H. J., Postoyan, R., Nešić, D. and Heemels, W. P. M. H. (2017). Computing Minimal and Maximal Allowable Transmission Intervals for Networked Control Systems Using the Hybrid Systems Approach, *IEEE Control Systems Letters*, 1, 1, 56-61, July.

Hespanha, J. P.; Naghshtabrizi, P.; Xu, Y. (2007). A Survey of Recent Results in Networked Control Systems, *IEEE Proceedings of the Technology of Networked Control Systems*, 95, 1, 138-162.

Lian, F. L.; Moyne, J. R; Tilbury, D. M. (2002). Network Design Consideration for Distributed Control Systems. *IEEE Transactions on Control Systems Technology*, 10(2), 297-307.

Martí, P.; Camacho, A.; Velasco, M.; Gaid, M. E. M. B. (2010). Runtime Allocation of Optional Control Jobs to a Set of CAN-Based Networked Control Systems. *IEEE Transactions on Industrial Informatics*, 6, 4, 503-520.

Martins, E. C., Jota, F. G. (2010). Design of Networked Control Systems with Explicit Compensation for Time-Delay Variations, *IEEE*

Transactions on Systems, Man and Cybernetics Part C: Applications and Reviews, 40, 3, 308-318.

Morawski M., Ignaciuk P. (2016) Adjustable Sampling Rate – An Efficient Way to Reduce the Impact of Network-Induced Uncertainty in Networked Control Systems?. In: Gaj P., Kwiecień A., Stera P. (eds) Computer Networks. *Communications in Computer and Information Science,* vol. 608. Springer.

Moyne, J. R.; Tilbury, D. M. (2007). The Emergence of Industrial Control Networks for Manufacturing Control, Diagnostics, and Safety Data, *IEEE Proceedings of the Technology of Networked Control Systems*, 95, 1, 29-47.

Penna, G. D., Intrigila, B., Magazzeni, D. (2009). Evaluating Fuzzy Controller Robustness Using Model Checking, Fuzzy Logic and Applications, *Lecture Notes in Computer Science*, V. D. Gesù, S. Pal, A. Petrosino, Eds. 5571, 303-311.

Pohjola, M. (2006). *PID Controller Design in Networked Control Systems*, Master's Thesis, Department of Automation and Systems Technology, Helsinki University of Technology.

In: Networked Control Systems
Editor: Eduardo Paciência Godoy
ISBN: 978-1-53613-105-5
© 2018 Nova Science Publishers, Inc.

Chapter 2

SWITCHING CONTROLLER SYNTHESIS FOR NETWORKED CONTROL SYSTEMS WITH VARYING SAMPLING INTERVALS

Vitor Mateus Moraes*[1,*]*, Eugênio de Bona Castelan Neto*[2]*,
***Ubirajara Franco Moreno*[2] *and Luis Almeida*[3]**
[1]Instituto Federal Catarinense – Campus São Francisco do Sul, São Francisco do Sul, Brazil
[2]Departamento de Automação e Sistemas, Centro Tecnológico, Universidade Federal de Santa Catarina, Florianópolis, Brazil
[3]Instituto de Telecomunicações, Faculdade de Engenharia da Universidade do Porto, Porto, Portugal

Abstract

Networked control systems pertain to a special class of sampled-data systems in which feedback control is performed over a communication network. When a communication network is shared among several systems, variable communication delay may occur, potentially affecting the control performance. Network restraints might be imposed as well, such as limitations on bandwidth usage, which also can affect the control performance, once it affects the rate at which the data from the systems are sampled. On the other hand, the control can be designed to tolerate a certain delay and its variation. According to that, we propose a compensator design method which takes into account both, delay and sampling

*E-mail address: vitor.moraes@ifc.edu.br.

intervals variations, guaranteeing stability and a certain performance level under some communication delay and bandwidth constraints, allowing a direct regulation of the amount of network resource to be used by a control system. Based on parameter dependent Lyapunov functions, we provide a closed-loop stability condition that can be verified, in terms of feasibility, through a set of linear matrix inequalities. A switching control synthesis is then promptly derived from the stability condition, leading to delay-dependent compensators. Furthermore, we present two numerical examples to demonstrate the use of the proposed methodology, one of which considers a scenario with multiple control tasks that may compete for a time-varying bandwidth.

Keywords: networked control system, parameter-dependent output feedback, switching control, co-design strategy

Introduction

Significant technological improvements, in terms of communication networks, have lead to an increasing usage of Networked Control Systems (NCSs) in industry, and consequently, a growing interest from the academic community (see, for instance, the survey papers Heemels and van de Wouw (2010); Zhang et al. (2013, 2016) and references therein). In such systems, the feedback control loop is closed over a communication network, typically shared among several control systems, which means that the control data are transmitted from one device to another through a communication network. This feature leads to some of the main advantages arising from the use of NCSs, such as ease of deployment and maintenance, flexibility, lower costs and increased integration of the information flows (Yang, 2006; Moyne and Tilbury, 2007).

Although the aforementioned benefits largely justify the usage of NCSs, the introduction of a communication network in the control loop may also bring some undesired effects, such as time-varying sampling intervals and delays, and even data losses. These effects may cause performance degradation, potentially leading to stability loss, thus rendering the stability analysis and the control design even more challenging (Tang and Yu, 2007). Thus, it is important to develop control techniques that cope with network induced timing features, such as time-varying delays (Moyne and Tilbury, 2007). Complementarily, an additional issue is that, in practice, usually not all plant states are measurable, thus the controller may only have access to the plant output values.

Under the above perspective, several methods of output feedback control synthesis for NCSs may be found in the literature. For instance, designing procedures using state-observers, which are mostly based on continuous-time modeling of the plant and the controller (Montestruque and Antsaklis, 2002; Naghshtabrizi and Hespanha, 2005). Alternatively, static feedback gains have also been proposed, as in Weihua and Minrui (2009), based on an uncertain continuous-time modeling. In Zhang et al. (2014), the authors presented a discrete-time approach with an approximation for delay compensation using a set of pre-computed gains. In Dritsas and Tzes (2007) an approach based on a discretized model of the continuous-time plant was used, taking into consideration an uncertain delay, although the control design was performed for a constant delay situation.

In particular, dynamic output feedback controllers for NCSs have been investigated mostly assuming either discrete-time plant modeling (Shi and Yu, 2011; Rasool et al., 2012; Mahmoud and Khan, 2015) or emulation approaches (Gao et al., 2010; Jiang et al., 2010), with some of them assuming quantized controllers. Still, it may be of relevant interest to have a more *sampled-data*-like approach, in the sense that the system behavior should be studied considering the digital control system interactions with a continuous-time plant. In this direction, Donkers et al. (2009) addressed a stability analysis using an approach based on an exact discretization of the plant, but no control design was provided. Also based on an exact discretization of the plant, Moraes et al. (2013) proposed an output feedback control design, taking into account some delay information.

Furthermore, in the context of NCSs, the available bandwidth may vary, due to the need to conserve energy, online dynamics adjustments or even resource overloads, which may lead to time-varying sampling intervals (Bouyssounouse and Sifakis, 2005). In any case, it is important to note that distinct sampling intervals correspond to distinct amounts of network bandwidth required by a control system. In Antunes et al. (2007), the authors presented a dynamic sampling rate adaptation, coordinated by a system manager: online resource allocation is carried out in order to prevent overload situations, which can increase the sampling intervals of the control-loops to increase available network bandwidth. In addition, the proposed strategy also used adaptive pole-placement controllers, but a thorough switching sampling interval stability analysis was not provided. Velasco et al. (2004) also presented an online bandwidth allocation strategy in which multiple control loops compete for network access. The authors proposed augmenting the original state-space representation of each controlled process with a new state variable describing the network usage dynamic, i.e., the assigned bandwidth to the control loop. A state-feedback control law was then designed taking into account the augmented state-space representation, although without explicitly considering the network induced delays. Recently, in Donkers et al. (2014), the authors presented a method for an output-based controller synthesis, taking into account time-varying sampling intervals and delays, considering both of them as norm-bounded uncertainties in the discretization modeling of the networked control system. However, a direct regulation of the amount of network resource to be used by a control system was not possible.

Aiming at considering the regulation of network resources, in this manuscript we present an output feedback control synthesis method that considers both the time-varying delays, and the capability of a closed-loop system to operate at several sampling intervals with arbitrary interchangeability between them. We also guarantee asymptotic stability by design, not only for each control subsystem operating mode individually, but also for any possible controller transition. We build on top of the delay-dependent compensator design method presented by Moraes et al. (2013), which already tunes the controller at run-time according to the current delay information. For simplicity, we make use of time-stamped messages, allowing the computation of the time-varying delays between sampling and actuation instants. Nevertheless, it is also possible to use delay estimates, as in Jungers et al. (2013), with minor modifications. The result is an adaptive output feedback compensator that takes into account both the delay and the bandwidth (sampling interval) variations at run time, guaranteeing the closed-loop performance requirements over all sampling interval transitions.

The remainder of the manuscript is organized as follows. The considered switching networked control system modeling and associated problem is described in the next section. In the sequence, switching closed-loop stability results are addressed, initially by presenting a stability condition based on a candidate switching parameter dependent Lyapunov function, followed by establishing a switching control synthesis, both written in terms of linear matrix inequalities (LMIs). Next, a section with numerical examples, describing two scenarios, is presented. The paper ends with some concluding remarks.

Notation

The following notational conventions are used throughout the manuscript. $\mathcal{N}$ stands for the set of natural numbers. $\Re$ stands for the set of real numbers. $\Re^n$

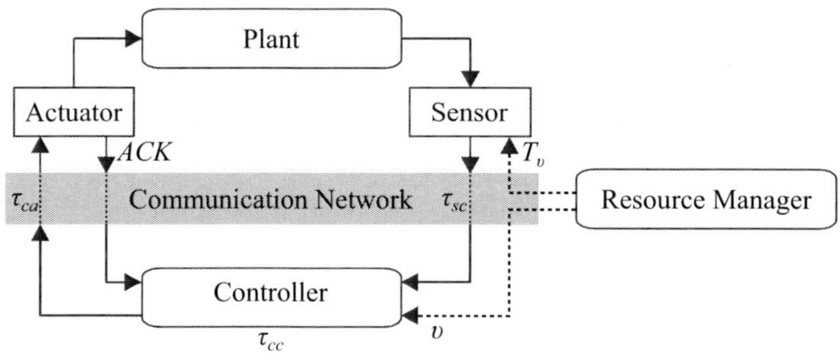

Figure 1. NCS with shared medium access controlled by a Resource Manager.

stands for the set of n-dimensional real vector space. $\Re^{n\times m}$ is the set of $n \times m$ real matrices. I (0) denotes an identity (zero) matrix with appropriate dimension, while I_s (0_s) is used to represent an s-dimensional identity (zero) matrix when necessary. For two symmetric matrices, A and B, $A > B$ means that $A - B$ is positive definite. By A' we denote the transpose of A and $He(A) = A + A'$. Also, $diag\{A, B\} = \left[\begin{smallmatrix} A & 0 \\ 0 & B \end{smallmatrix}\right]$. The symbol $*$ stands for symmetric blocks, while $(\bullet)$ stands for an element that has no influence on the development. The Euclidian norm is represented by $\|\cdot\|$.

Problem Description

Let us consider an NCS as depicted in Figure 1, where the sensor is considered to be time-driven at regular time intervals that may change among a pre-specified set of sampling intervals. Apart this switching feature, the behaviour of the NCS is also dictated by the following (see Moraes et al. (2013); Hetel et al. (2006)): the message sent from the sensor to the controller is time-stamped; the controller and the actuator are event-driven, with the latter acting as a zero-order-holder and also sending a time-stamped message to the controller, denoted ACK, containing the information about the instant when the control signal is applied to the plant. In this NCS scheme, the global time-varying delay between sampling and actuation is given by $\tau = \tau_{sc} + \tau_{cc} + \tau_{ca}$, where: τ_{sc} is the sensor-to-controller delay, τ_{cc} is the computation delay and τ_{sc} is the controller-to-actuator delay. Thus, by considering this operation scheme and the use of reliable network environments with deterministic medium access control protocols[1], the time-stamped messages allow the use of the actual delay information, about the current discrete-time interval, to compute a control signal to be applied on the next sampling interval, as presented in the following.

The plant is described by a continuous linear time-invariant system:

$$\mathscr{P} : \begin{cases} \dot{x}(t) = Mx(t) + Nu(t) \\ y(t) = Cx(t) \end{cases} \tag{1}$$

with $x(t) \in \Re^n$, $u(t) \in \Re^m$, $y(t) \in \Re^p$, $M \in \Re^{n\times n}$, $N \in \Re^{n\times m}$ and $C \in \Re^{p\times n}$, with $p < n$.

To represent the switching dynamics of the plant in discrete-time, with respect to the sampling instants kT_υ, $k \in \mathscr{N}$, consider that the sampling intervals belong to a finite ordered set $\{T_\upsilon;\ \upsilon = 1,\ldots,N_\upsilon\}$, such that $0 < T_1 \leq T_2 \leq \cdots \leq T_{N_\upsilon}$. The index υ stands for the current sensor sampling interval of the plant, and it can assume any value in $\{T_\upsilon\}$ at any arbitrary instant k, thus defining the operating mode of the switching system. Additionally, due to the assumption that real-time features are provided by the communication network and medium access control protocol (Wittenmark et al., 2002), we can consider the following delay bounds for each particular sampling interval, all being dependent on the operating mode: $\tau \in [\tau_{\upsilon,\min},\ \tau_{\upsilon,\max}]\ \forall k$, with $0 < \tau_{\upsilon,\min} < \tau_{\upsilon,\max} \leq T_\upsilon$. Herein, τ is the delay and $\tau_{\upsilon,\min}$ and $\tau_{\upsilon,\max}$ are, respectively, the minimum and maximum delay when the sampling interval is T_υ.

The control signal applied to the plant, over two consecutive sampling instants, is given by[2]:

$$u(t) = \begin{cases} u^-, & t \in [kT_\upsilon,\ kT_\upsilon + \tau) \\ u, & t \in [kT_\upsilon + \tau,\ (k+1)T_\upsilon) \end{cases} \tag{2}$$

which, by applying an exact discretization over the same interval, leads to:

$$\mathscr{P}_\upsilon : \begin{cases} x^+ = A_\upsilon x + \tilde{\Gamma}_\upsilon(\tau)u^- + \Gamma_\upsilon(\tau)u \\ y = Cx \end{cases} \tag{3}$$

[1]Under these network conditions, packet losses are not likely to happen and a maximum message delivery time can be guaranteed through a correct scheduling of the shared medium access.

[2]For ease of notation, the index k is omitted, while the indices $k+1$ and $k-1$ are replaced by a superscript $+$ and $-$, respectively.

$\forall \upsilon = 1,\ldots,N_\upsilon$, with dynamic matrices $A_\upsilon = e^{MT_\upsilon}$, $B_\upsilon = \int_0^{T_\upsilon} e^{Ms} ds\, N$,

$$\begin{aligned} \Gamma_\upsilon(\tau) &= \int_0^{T_\upsilon - \tau} e^{Ms} ds\, N, \\ \tilde{\Gamma}_\upsilon(\tau) &= \int_{T_\upsilon - \tau}^{T_\upsilon} e^{Ms} ds\, N = B_\upsilon - \Gamma_\upsilon(\tau). \end{aligned} \tag{4}$$

Considering an extended state vector $\bar{x} = [x' \; u'^{-} \; u']' \in \Re^l$, $l = n + 2m$, we can rewrite the discrete-time system as:

$$\begin{aligned} \bar{x}^+ &= \bar{A}_\upsilon(\tau)\bar{x} + \bar{B}u^+ \\ y &= \bar{C}\bar{x} \end{aligned} \tag{5}$$

where $\bar{B} = \begin{bmatrix} 0_{m\times n} & 0_m & I_m \end{bmatrix}'$, $\bar{C} = \begin{bmatrix} C & 0_{p\times m} & 0_{p\times m} \end{bmatrix}$,

$$\bar{A}_\upsilon(\tau) = \begin{bmatrix} A_\upsilon & B_\upsilon - \Gamma_\upsilon(\tau) & \Gamma_\upsilon(\tau) \\ 0_{m\times n} & 0_m & I_m \\ 0_{m\times n} & 0_m & 0_m \end{bmatrix}. \tag{6}$$

Notice that the actual information about the time-varying parameter τ is computed only after reception by the controller of the *ACK* message sent by the actuator. This requires considering a control input delay that is reflected in the used formulation by the presence of u in the augmented state-vector $\bar{x}$ and by defining u^+ as the control input.

Since (5) has parameter-dependent matrices, with respect to the operating mode T_υ and the time-varying delay τ, for each operating mode, a polytopic representation with additional norm-bounded uncertainty is used, based on an approximation using a Taylor series expansion (Hetel et al., 2006; Moraes et al., 2013):

$$\Gamma_\upsilon(\tau) = \sum_{i=1}^{N_{h_\upsilon}} \mu_{\upsilon,i}(\tau)\Gamma_{\upsilon,i} + \Delta_\upsilon(\tau), \tag{7}$$

with $\sum_{i=1}^{N_{h_\upsilon}} \mu_{\upsilon,i}(\tau) = 1$, $\mu_{\upsilon,i}(\tau) \geq 0$, $N_{h_\upsilon} = h_\upsilon + 1$, where h_υ is the approximation order of the Taylor series expansion related to the system dynamics over the sampling interval T_υ.

The polytope vertices are given by:

$$\Gamma_{\upsilon,i} = \begin{bmatrix} \frac{M^{h_\upsilon - 1}}{h_\upsilon !} & \cdots & \frac{M}{2!} & I \end{bmatrix} \phi_{\upsilon,i} N, \tag{8}$$

$\Gamma_{\upsilon,i} \in \Re^{n\times m}$, $\phi_{\upsilon,i} \in \Re^{h_\upsilon n\times n}$, $\forall i = 1, \ldots, h_\upsilon + 1$, where:

$$\phi_{\upsilon,1} = \begin{bmatrix} \underline{\alpha}_\upsilon^{h_\upsilon} I \\ \vdots \\ \underline{\alpha}_\upsilon^2 I \\ \underline{\alpha}_\upsilon I \end{bmatrix}, \quad \phi_{\upsilon,2} = \begin{bmatrix} \underline{\alpha}_\upsilon^{h_\upsilon} I \\ \vdots \\ \underline{\alpha}_\upsilon^2 I \\ \bar{\alpha}_\upsilon I \end{bmatrix}, \quad \ldots, \quad \phi_{\upsilon,h_\upsilon+1} = \begin{bmatrix} \bar{\alpha}_\upsilon^{h_\upsilon} I \\ \vdots \\ \bar{\alpha}_\upsilon^2 I \\ \bar{\alpha}_\upsilon I \end{bmatrix}, \tag{9}$$

with $\underline{\alpha}_\upsilon = T_\upsilon - \tau_{\upsilon,\max}$ and $\bar{\alpha}_\upsilon = T_\upsilon - \tau_{\upsilon,\min}$.

In addition, since the residual uncertainty $\Delta(\tau_k) \in \Re^{n\times m}$ is considered to be norm-bounded, i.e.:

$$\left\|\Delta_\upsilon(\tau)\right\| \leq \gamma_\upsilon, \tag{10}$$

an upper bound of γ_υ, denoted by $\bar{\gamma}_\upsilon$, can be estimated *offline* by applying a gridding approach.

Thus, considering any arbitrary switching rule for system (5), i.e., any arbitrary change among the possible sampling intervals, the following problem is proposed.

Problem 1. *Design dynamic output feedback compensators with full-order $l = n+2m$, given by:*

$$\mathscr{K}_\upsilon : \begin{cases} \zeta^+ = \mathscr{A}_\upsilon(\tau)\zeta + \mathscr{B}_\upsilon(\tau)y \\ u^+ = \mathscr{C}_\upsilon(\tau)\zeta + \mathscr{D}_\upsilon(\tau)y \end{cases} \tag{11}$$

$\upsilon \in \{1,\ldots,N_\upsilon\}$, *such that the switching closed-loop system is asymptotically stable.*

In addition, the delay-dependent compensator matrices may also be structured as polytopes, i.e.,

$$\begin{bmatrix} \mathscr{A}_\upsilon(\tau) & \mathscr{B}_\upsilon(\tau) \\ \mathscr{C}_\upsilon(\tau) & \mathscr{D}_\upsilon(\tau) \end{bmatrix} = \sum_{i=1}^{N_{h_\upsilon}} \mu_{\upsilon,i}(\tau) \begin{bmatrix} \mathscr{A}_{\upsilon,i} & \mathscr{B}_{\upsilon,i} \\ \mathscr{C}_{\upsilon,i} & \mathscr{D}_{\upsilon,i} \end{bmatrix}. \tag{12}$$

Remark 1. *The* online *computations necessary to implement the compensator $\mathscr{K}_\upsilon$ concern the determination of the varying delay τ, the calculation of weights $\mu_{\upsilon,i}(\tau)$ (for details, see Moraes et al. (2013)), and the computation of the controller output u^+ from (11-12). Thus, the vertex matrices $\mathscr{A}_{\upsilon,i}$, $\mathscr{B}_{\upsilon,i}$, $\mathscr{C}_{\upsilon,i}$ and $\mathscr{D}_{\upsilon,i}$, which are computed* offline, *have to be available in the processor memory.*

Remark 2. *The computation delay τ_{cc} is related to the numerical complexity involved in the above mentioned* online *computations. Thus, it is a function of the compensator order, fixed as $l = n+2m$; the assigned number of operating modes, N_υ; and the number of vertices involved in the polytopic model, N_{h_υ}, which turns out to be determined by the designer's choice of Taylor Series order truncation h_υ.*

Remark 3. *Considering an NCS context similar to the one described in this section, but assuming a constant sampling interval T, the dynamic output feedback control of a plant represented by* (1) *was treated in Moraes et al. (2013) using a parameter-varying but non-switching compensator $\mathscr{K}$. Thus, by considering a single operating mode ($\upsilon = 1$), the results proposed by Moraes et al. (2013) represent a special case of the results developed in the present work.*

Switching Closed-Loop Stability

Let us define the auxiliary variable $z = [\bar{x}' \;\; \zeta']' \in \Re^{2l}$. Then, a switching closed-loop representation is given by:

$$z^+ = H_\upsilon(\tau)z + E\Delta_\upsilon(\tau)Dz, \tag{13}$$

where $E = \begin{bmatrix} I_n & 0_{n\times m} & 0_{n\times m} & 0_{n\times l}\end{bmatrix}'$, $D = \begin{bmatrix} 0_{m\times n} & -I_m & I_m & 0_{m\times l}\end{bmatrix}$, $H_\upsilon(\tau) = \sum_{i=1}^{N_{h_\upsilon}} \mu_{\upsilon,i}(\tau)H_{\upsilon,i}$, and

$$H_{\upsilon,i} = \begin{bmatrix} \bar{A}_{\upsilon,i} + \bar{B}\mathscr{D}_{\upsilon,i}\bar{C} & \bar{B}\mathscr{C}_{\upsilon,i} \\ \mathscr{B}_{\upsilon,i}\bar{C} & \mathscr{A}_{\upsilon,i} \end{bmatrix}, \tag{14}$$

with, from (6) and (7):

$$\bar{A}_{\upsilon,i} = \begin{bmatrix} A_\upsilon & B_\upsilon - \Gamma_{\upsilon,i} & \Gamma_{\upsilon,i} \\ 0_{m\times n} & 0_m & I_m \\ 0_{m\times n} & 0_m & 0_m \end{bmatrix}. \tag{15}$$

Definition 1. *Given a candidate switching parameter dependent Lyapunov function (SPDLF):*

$$V_\upsilon(z,\tau) = z'Q_\upsilon^{-1}(\tau)z, \tag{16}$$

with $Q_\upsilon(\tau) = \sum_{i=1}^{N_{h_\upsilon}} \mu_{\upsilon,i}(\tau)Q_{\upsilon,i}$, $Q_{\upsilon,i} = Q'_{\upsilon,i} > 0$ and $\sum_{i=1}^{N_{h_\upsilon}} \mu_{\upsilon,i}(\tau) = 1$, the switching closed-loop system is robustly asymptotically stable, with contractivity coefficient $\lambda \in (0,1]$, if:

$$\Delta V \triangleq V_{\upsilon^+}(z^+,\tau^+) - \lambda V_\upsilon(z,\tau) < 0, \tag{17}$$

$\forall z \in \Re^{2l}$, $z \neq 0$ *and* $\forall \tau \in [\tau_{\min}, \tau_{\max}]$.

Remark 4. *For the purpose of performance measurement, a constant contractivity coefficient* λ *is used in Definition 1, which leads to a global minimum decay rate* α *that is related to the maximum sampling interval,* T_{N_υ} *through the classical relation:*

$$\lambda = e^{-\alpha T_{N_\upsilon}} \quad \Longleftrightarrow \quad \alpha = -\frac{\ln \lambda}{T_{N_\upsilon}}. \tag{18}$$

This means that there is a performance improvement when smaller sampling intervals are used.

Applying Definition 1 to the switching closed-loop system (13), leads to:

$$\big(H_\upsilon(\tau) + E\Delta_\upsilon(\tau)D\big)' Q_{\upsilon^+}^{-1}(\tau^+)\big(H_\upsilon(\tau) + E\Delta_\upsilon(\tau)D\big) - \lambda Q_\upsilon^{-1}(\tau) < 0, \tag{19}$$

from which the following Lemma can be stated.

Lemma 1. *Consider a given* $\lambda \in (0,1]$ *and the upper bounds* $\bar{\gamma}_\upsilon$ *computed for each operating mode* $\upsilon \in \{1,\ldots,N_\upsilon\}$. *The switching closed-loop system* (13) *is robustly asymptotically stable if there exist symmetric positive definite matrices* $\tilde{Q}_{r,i} \in \Re^{2l\times 2l}$, *and matrices* $\tilde{U}_r \in \Re^{2l\times 2l}$ *that verify:*

$$\begin{bmatrix} -\tilde{Q}_{q,j} + \bar{\gamma}_r^2 EE' & H_{r,i}\tilde{U}_r & 0 \\ * & \lambda(\tilde{Q}_{r,i} - \tilde{U}_r - \tilde{U}_r') & \tilde{U}_r' D' \\ * & * & -I \end{bmatrix} < 0, \tag{20}$$

$\forall r, q = 1, \ldots, N_\upsilon$ *and* $\forall i, j = 1, \ldots, N_{h_\upsilon}$.

Proof. First, notice that the indices r and q are related to the current operating mode in the k-th and $(k+1)$-th instants, respectively. Thus, evaluating (20) gives

$$\begin{bmatrix} -\tilde{Q}_{\upsilon^+,j} + \bar{\gamma}_\upsilon^2 EE' & H_{\upsilon,i}\tilde{U}_\upsilon & 0 \\ * & \lambda(\tilde{Q}_{\upsilon,i} - \tilde{U}_\upsilon - \tilde{U}_\upsilon') & \tilde{U}_\upsilon' D' \\ * & * & -I \end{bmatrix} < 0, \tag{21}$$

$\forall i, j = 1, \ldots, N_{h_\upsilon}$.

Moreover, the indices i and j are related to the polytope vertices, also in the k-th and $(k+1)$-th instants, respectively. Thus, defining $Q_{r,i} = \sigma_r \tilde{Q}_{r,i}$ and $U_r = \sigma_r \tilde{U}_r$, with σ_r being any positive scalar, by pre- and post-multiplying (20) by $\sqrt{\sigma_r} I_{(4l+m)}$, appropriately performing convex combinations, first for i and afterward for j, and applying the Schur complement (Boyd et al., 1994), leads to:

$$\begin{bmatrix} -Q_{\upsilon^+}(\tau^+) + \sigma_\upsilon \bar{\gamma}_\upsilon^2 E E' & H_\upsilon(\tau) U_\upsilon \\ * & \lambda(Q_\upsilon(\tau) - U_\upsilon - U_\upsilon') + \sigma_\upsilon^{-1} U_\upsilon' D' D U_\upsilon \end{bmatrix} < 0. \quad (22)$$

This is equivalent to

$$\begin{bmatrix} -Q_{\upsilon^+}(\tau^+) & H_\upsilon(\tau) U_\upsilon \\ * & \lambda(Q_\upsilon(\tau) - U_\upsilon - U_\upsilon') \end{bmatrix} + \sigma_\upsilon \begin{bmatrix} \bar{\gamma}_\upsilon E \\ 0 \end{bmatrix} \begin{bmatrix} \bar{\gamma}_\upsilon E' & 0 \end{bmatrix} + \sigma_\upsilon^{-1} \begin{bmatrix} 0 \\ U_\upsilon' D' \end{bmatrix} \begin{bmatrix} 0 & D U_\upsilon \end{bmatrix} < 0. \quad (23)$$

Since $\Delta_\upsilon(\tau)' \Delta_\upsilon(\tau) \leq \bar{\gamma}_\upsilon^2 I$, and by the fact that $-U_\upsilon' Q_\upsilon^{-1}(\tau) U_\upsilon \leq Q_\upsilon(\tau) - U_\upsilon - U_\upsilon'$, the previous inequality implies

$$\begin{bmatrix} -Q_{\upsilon^+}(\tau^+) & (H_\upsilon(\tau) + E\Delta_\upsilon(\tau) D) \\ (H_\upsilon(\tau) + E\Delta_\upsilon(\tau) D)' & -\lambda Q_\upsilon^{-1}(\tau) \end{bmatrix} < 0, \quad (24)$$

that by the Schur complement leads to (19). □

Based on the stability condition established by Lemma 1, we can now derive a solution to the switching control synthesis problem.

Switching Control Synthesis

Aiming at synthesizing the proposed dynamic output feedback controller (11-12), motivated by the approach used in Castelan et al. (2010), which was inspired by Scherer et al. (1997), in the following, we consider the auxiliary matrices:

$$\tilde{U}_\upsilon = \begin{bmatrix} X_\upsilon & (\bullet) \\ Z_\upsilon & (\bullet) \end{bmatrix}, \quad \tilde{U}_\upsilon^{-1} = \begin{bmatrix} Y_\upsilon & (\bullet) \\ W_\upsilon & (\bullet) \end{bmatrix}, \quad \Theta_\upsilon = \begin{bmatrix} Y_\upsilon & I \\ W_\upsilon & 0 \end{bmatrix}, \quad (25)$$

that verify:

$$\Psi_\upsilon = \tilde{U}_\upsilon \Theta_\upsilon = \begin{bmatrix} I & X_\upsilon \\ 0 & Z_\upsilon \end{bmatrix}, \quad \hat{U}_\upsilon = \Theta_\upsilon' \tilde{U}_\upsilon \Theta_\upsilon = \begin{bmatrix} Y_\upsilon' & T_\upsilon' \\ I & X_\upsilon \end{bmatrix}, \quad (26)$$

where, by construction:

$$T_\upsilon' = Y_\upsilon' X_\upsilon + W_\upsilon' Z_\upsilon. \quad (27)$$

We also assume the change of variable $\hat{Q}_\upsilon(\tau) = \Theta_\upsilon' \tilde{Q}_\upsilon(\tau) \Theta_\upsilon$.

Proposition 1. *Consider a given $\lambda \in (0,1]$ and the upper bounds $\bar{\gamma}_\upsilon$ computed for each operating mode $\upsilon \in \{1,\ldots,N_\upsilon\}$. Consider the existence of symmetric positive definite matrices $\hat{Q}_{r,i}$, and matrices Y_r, X_r, T_r, $\hat{\mathscr{A}}_{r,i}$, $\hat{\mathscr{B}}_{r,i}$, $\hat{\mathscr{C}}_{r,i}$, $\hat{\mathscr{D}}_{r,i}$ that verify:*

$$\begin{bmatrix} -\hat{Q}_{q,j} & \Omega_{r,i} & 0 & \bar{\gamma}_r\Theta_r'E \\ * & \lambda(\hat{Q}_{r,i}-\hat{U}_r-\hat{U}_r') & \Psi_r'D' & 0 \\ * & * & -I & 0 \\ * & * & * & -I \end{bmatrix} < 0, \tag{28}$$

$\forall r,q = 1,...,N_\upsilon$ and $\forall i,j = 1,...,N_{h_\upsilon}$, and

$$\Omega_{r,i} = \begin{bmatrix} Y_r'\bar{A}_{r,i}+\hat{\mathscr{B}}_{r,i}\bar{C} & \hat{\mathscr{A}}_{r,i} \\ \bar{A}_{r,i}+\bar{B}\hat{\mathscr{D}}_{r,i}\bar{C} & \bar{A}_{r,i}X_r+\bar{B}\hat{\mathscr{C}}_{r,i} \end{bmatrix}. \tag{29}$$

Let W_r and Z_r be any nonsingular matrices such that

$$W_r'Z_r = T_r' - Y_r'X_r. \tag{30}$$

Then the dynamic output feedback compensators (11) *with*

$$\begin{cases} \mathscr{D}_{r,i} = \hat{\mathscr{D}}_{r,i} \\ \mathscr{C}_{r,i} = (\hat{\mathscr{C}}_{r,i}-\mathscr{D}_{r,i}\bar{C}X_r)Z_r^{-1} \\ \mathscr{B}_{r,i} = (W_r')^{-1}(\hat{\mathscr{B}}_{r,i}-Y_r'\bar{B}\mathscr{D}_{r,i}) \\ \mathscr{A}_{r,i} = (W_r')^{-1}(\hat{\mathscr{A}}_{r,i}-Y_r'(\bar{A}_{r,i}+\bar{B}\mathscr{D}_{r,i}\bar{C})X_r-Y_r'\bar{B}\mathscr{C}_{r,i}Z_r-W_r'\mathscr{B}_{r,i}\bar{C}X_r)Z_r^{-1} \end{cases} \tag{31}$$

are such that the switching closed-loop system (13) *is robustly asymptotically stable, thus solving Problem 1.*

Proof. By applying the Schur complement to the upper left term of (20), we have:

$$\begin{bmatrix} -\tilde{Q}_{q,j}' & H_{r,i}\tilde{U}_r & 0 & \bar{\gamma}_rE \\ * & \lambda(\tilde{Q}_{r,i}-\tilde{U}_r-\tilde{U}_r') & \tilde{U}_r'D' & 0 \\ * & * & -I & 0 \\ * & 0 & 0 & -I \end{bmatrix} < 0, \tag{32}$$

Thus, by pre- and post-multiplying this inequality by $diag\{\Theta_\upsilon',\Theta_\upsilon',I,I\}$ and its transpose, respectively, and defining the auxiliary variables:

$$\begin{cases} \hat{\mathscr{D}}_{r,i} = \mathscr{D}_{r,i} \\ \hat{\mathscr{C}}_{r,i} = \mathscr{D}_{r,i}\bar{C}X_r + \mathscr{C}_{r,i}Z_r \\ \hat{\mathscr{B}}_{r,i} = Y_r'\bar{B}\mathscr{D}_{r,i} + W_r'\mathscr{B}_{r,i} \\ \hat{\mathscr{A}}_{r,i} = Y_r'(\bar{A}_{r,i} + \bar{B}\mathscr{D}_{r,i}\bar{C})X_r + Y_r'\bar{B}\mathscr{C}_{r,i}Z_r + W_r'\mathscr{B}_{r,i}\bar{C}X_r + W_r'\mathscr{A}_{r,i}Z_r \end{cases} \tag{33}$$

the equivalence between (28) and (20) is set. Thus, from Lemma 1, the corresponding switching closed-loop system (13) is robustly asymptotically stable. □

Disregarding that this synthesis result has been presented with respect to the control of a single plant, it can be used for co-designing purposes. For instance, in a networked control of several plants as depicted in Figure 6, where each plant is subject to two or more sampling intervals that may be arbitrarily assigned by an external supervisory process, namely a Resource Manager (RM) (for instance, see Antunes et al. (2007); Velasco et al. (2004)), which controls the bandwidth resources needed by the different tasks that share the network medium, including the control tasks. This is one of the scenarios exploited in the following session.

Illustrative Example

To illustrate the use of the proposed results, two different scenarios are now considered. In the first one, aiming at showing some consequences of switching between sampling intervals on NCS performance, a single control task subject to arbitrary switching is considered. Afterward, in the second scenario, multiple control tasks compete for the time-varying bandwidth by means of an RM, which leads to arbitrary interchangeability of sampling intervals. The presented simulations were performed using the *TrueTime* toolbox (Cervin et al., 2003)[3].

---

[3]Complementary numerical data can be obtained from the authors.

Switching Controller

For this example, the following continuous linear time-invariant system presented in Donkers et al. (2009) is considered:

$$M = \begin{bmatrix} 1.380 & -0.208 & 6.715 & -5.676 \\ -0.581 & -4.290 & 0 & 0.675 \\ 1.067 & 4.273 & -6.654 & 5.893 \\ 0.048 & 4.273 & 1.343 & -2.104 \end{bmatrix}, N = \begin{bmatrix} 0 & 0 \\ 5.679 & 0 \\ 1.136 & -3.146 \\ 1.136 & 0 \end{bmatrix}, C' = \begin{bmatrix} 1 & 0 \\ 0 & 1 \\ 1 & 0 \\ -1 & 0 \end{bmatrix}, \tag{34}$$

which has the open-loop eigenvalues: 1.99, 0.0585, −5.0484 and −8.6681.

To show the consequences of switching between sampling intervals on an NCS, two operating modes are assumed, tuned considering some arbitrary desired performance for each. The sampling intervals are $T_1 = 10$ms and $T_2 = 60$ms. For both operating modes, the approximation order of the TSA representation is $h = 2$ and the minimum communication delay bound is $\tau_{\min} = 0$, whereas the maximum is always equal to the sampling interval that is being used. In addition, the communication delay is randomly generated between the aforementioned bounds.

Initially, choosing $\lambda = 0.9$, based on the proposed compensator synthesis and following Remark 3, it is possible to individually design two compensators, both leading to stable closed-loop systems when running full-time without switching, as shown in Figure 2 (i.e., for each situation shown in the figure), over the entire simulation interval, there is only one compensator being used. However, when these two synthesized compensators are implemented in a switching fashion, arbitrarily alternating between both operation modes (as shown in Figure 4), the closed-loop system does not perform properly, eventually leading to unstable behavior as seen in Figure 3.

On the other hand, when applying the switching design from Proposition 1, a feasible solution is reached for the same set of operating modes $T_1 = 10$ms and $T_2 = 60$ms. Moreover, the obtained compensators may be arbitrarily switched, along with the sampling intervals, preserving the closed-loop stability and relative performance given by $\lambda = 0.9$, as can be seen in Figure 5 for the same switching sampling intervals sequence used before.

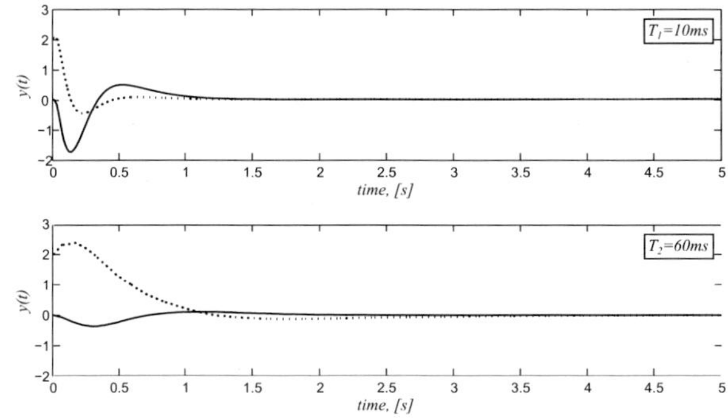

Figure 2. System outputs, $y_1(t)$ $\cdots$ and $y_2(t)$ —, for $T_1 = 10$ms and $T_2 = 60$ms.

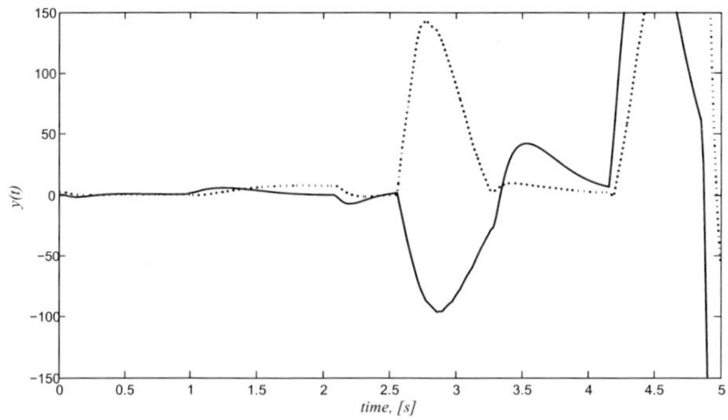

Figure 3. System outputs, $y_1(t)$ $\cdots$ and $y_2(t)$ —, with switching sampling intervals and not considering the proposed switching control design.

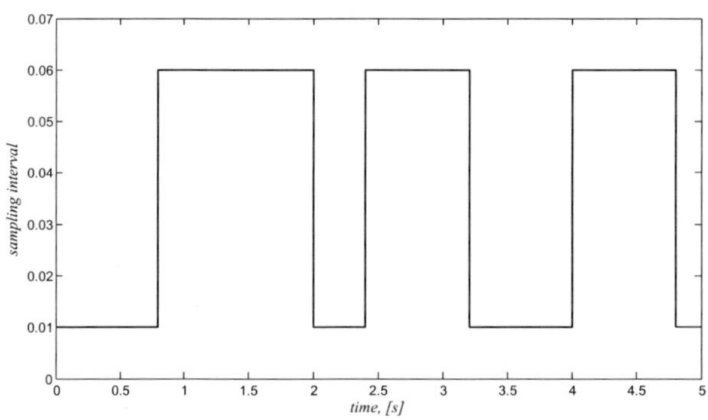

Figure 4. Sampling interval over time, with $T_1 = 10$ms and $T_2 = 60$ms.

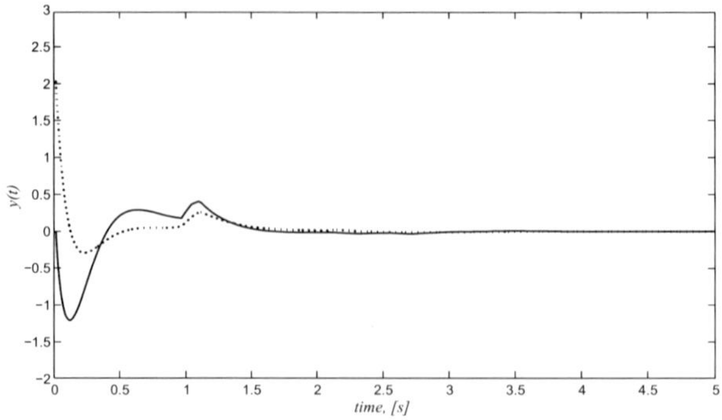

Figure 5. System outputs, $y_1(t)$ $\cdots$ and $y_2(t)$ —, with switching sampling intervals and considering the proposed design.

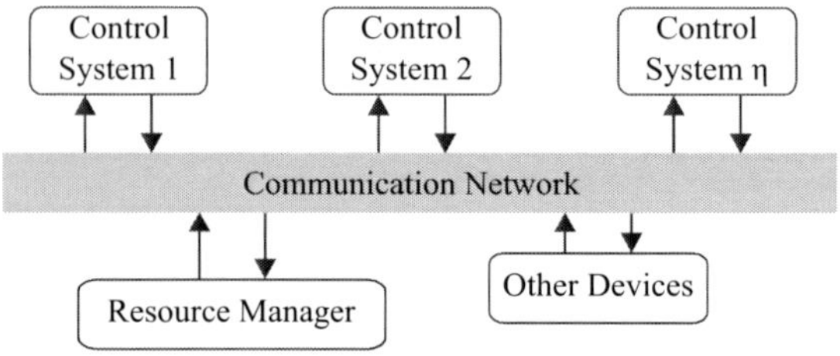

Figure 6. Multiple tasks sharing a network with a resource manager controlling the bandwidth usage.

Multiple Controller Tasks

Now, we consider a more comprehensive networked control system, in the sense that multiple tasks may compete for a shared network resource, as depicted in Figure 6. In this situation, since distinct amounts of bandwidth may imply distinct sampling intervals, once a task has been granted a larger amount of the resource, the other tasks shall have their sampling intervals adapted accordingly, preserving the reliability of the network usage.

Although the approach proposed here can be used to deal with an arbitrary fixed number of sampling intervals, we assume in this example that each control loop has only two operating modes. Additionally, we assume the existence of an RM, which has the information of the network bandwidth utilization level of all tasks and is allowed to perform a re-scheduling policy once every interval

T_{rm}, hence coordinating the switching between operating modes of each control system.

In this second illustrative example, a set of three continuous linear time-invariant systems is considered, all described by continuous linear time-invariant dynamics:

$$\mathscr{P}_\ell : \begin{cases} \dot{x}_\ell(t) = M_\ell x_\ell(t) + N_\ell u_\ell(t) \\ y_\ell(t) = C_\ell x_\ell(t) \end{cases} \tag{35}$$

with $\ell \in \{1,2,3\}$.

In addition to the system used in the previous example, denoted by M_2, the other two systems added to the NCS are a second-order plant with:

$$M_1 = \begin{bmatrix} 0 & 1 \\ 288.24 & 0 \end{bmatrix}, \; N_1 = \begin{bmatrix} 0 \\ -73.53 \end{bmatrix}, \; C_1 = \begin{bmatrix} 1 & 0 \end{bmatrix}, \tag{36}$$

which has the open-loop eigenvalues: 16.9775 and -16.9775, and a first-order plant with:

$$M_3 = \frac{5}{100}, \; N_3 = \frac{1}{1000}, \; C_3 = 1. \tag{37}$$

A CAN network is used for simulation, with an adaptive rate monotonic medium access priority assignment, which is dependent on the current sampling interval from each task. The data transfer speed is 250Kbps, which for a packet size of 108bits and computation time $\tau_{cc} = 0.5$ms gives a minimum delay $\tau_{\min} \approx 1.4$ms. The maximum delay is always assumed to be equal to the current sampling interval, which is guaranteed by the message scheduling. Furthermore, for the online simulation, the access to the network is randomly disturbed by some other devices, inducing time-varying delays, and it is assumed that the Resource Manager runs once every $T_{rm} = 0.5$s. Observe that, since there is an RM supervising the network bandwidth allocation, the delays always remain between the previously mentioned bounds.

For each system, after some tunning looking for a suitable individual performance, the following nominal contractivity coefficients and sampling intervals are set: $\lambda_1 = 0.95$, $T_{1,1} = 8$ms, $T_{1,2} = 24$ms; $\lambda_2 = 0.90$, $T_{2,1} = 10$ms, $T_{2,2} = 60$ms; and $\lambda_3 = 0.96$, $T_{3,1} = 15$ms, $T_{3,2} = 60$ms. Notice that the contractivity coefficient is the same for both operating modes associated to each plant.

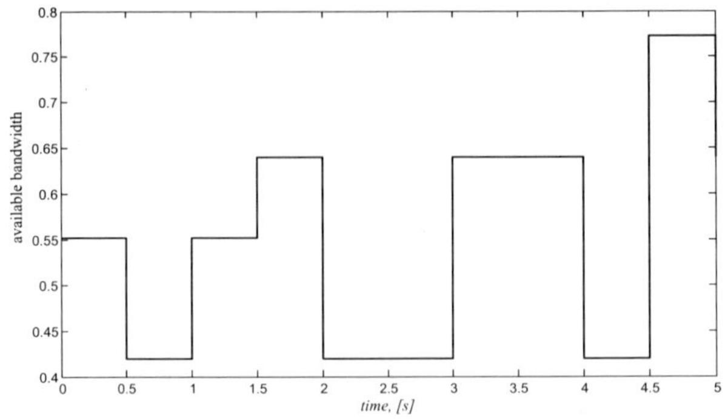

Figure 7. Available bandwidth over time.

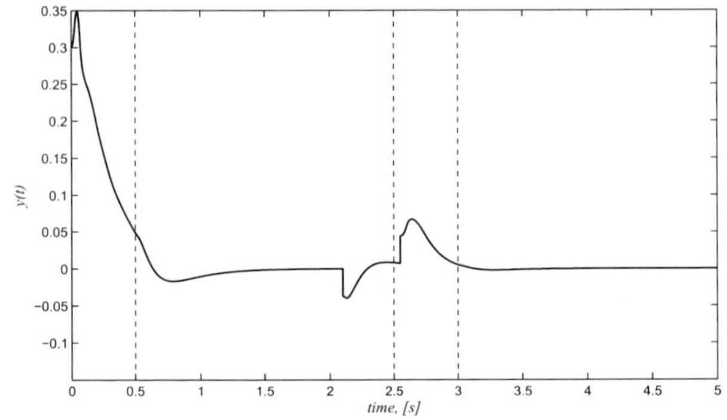

Figure 8. Plant 1 output.

The simulation runs for 5s and, over the interval, the network bandwidth that is available to the control systems varies according to the plot shown in Figure 7, which complies with an initial setup: $T_{1,2}$, $T_{2,1}$ and $T_{3,2}$.

At $t = 2$s, some perturbation are applied to the states of the systems: a square pulse to systems 1 and 2, and a step to plant 3. Observe that only at 4.5s, when a larger amount of bandwidth become available for the use of the control systems, it is possible for all three to run at the shortest sampling interval, $T_{i,1}$, $i = 1,2,3$, as shown in figures 9, 11 and 13. As depicted in Figure 8, Figure 10 and Figure 12, all closed-loop systems remain stable and reach equilibrium in any situation. Notice that in these figures the dashed vertical lines indicate when the sampling intervals were switched, for each system.

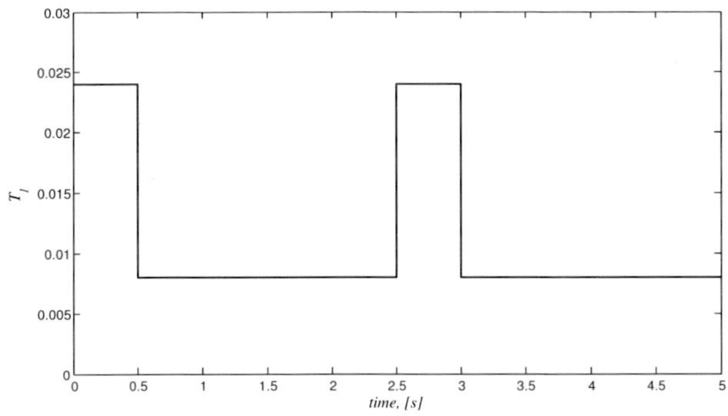

Figure 9. Plant 1 sampling interval over time, with $T_{1,1} = 8$ms and $T_{1,2} = 24$ms.

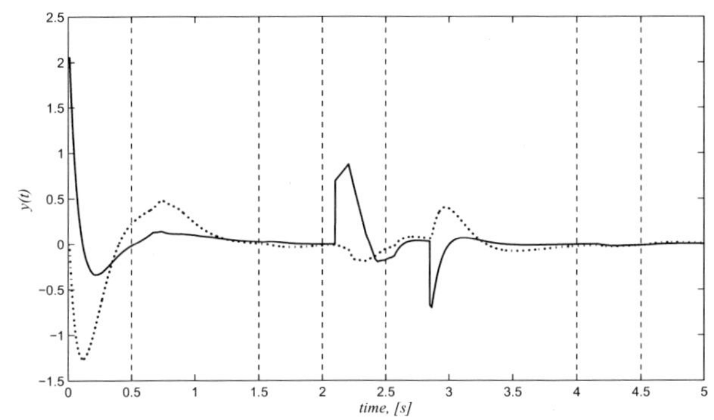

Figure 10. Plant 2 outputs, $y_1(t)$ — and $y_2(t)$ $\cdots$.

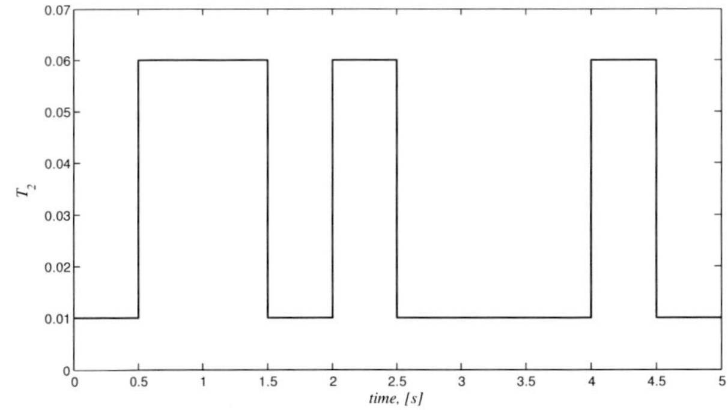

Figure 11. Plant 2 sampling interval over time, with $T_{2,1} = 10$ms and $T_{2,2} = 60$ms.

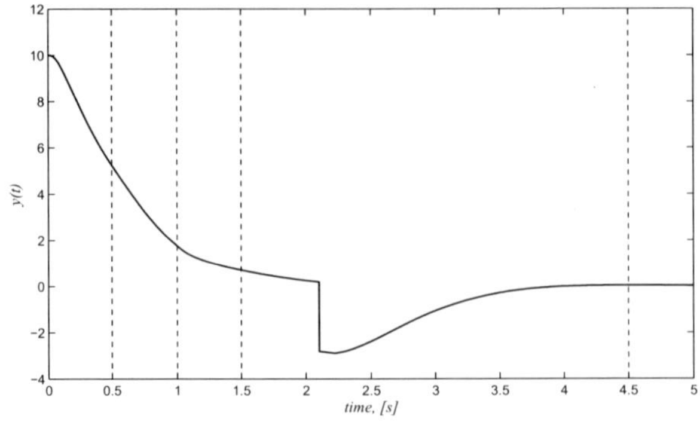

Figure 12. Plant 3 output.

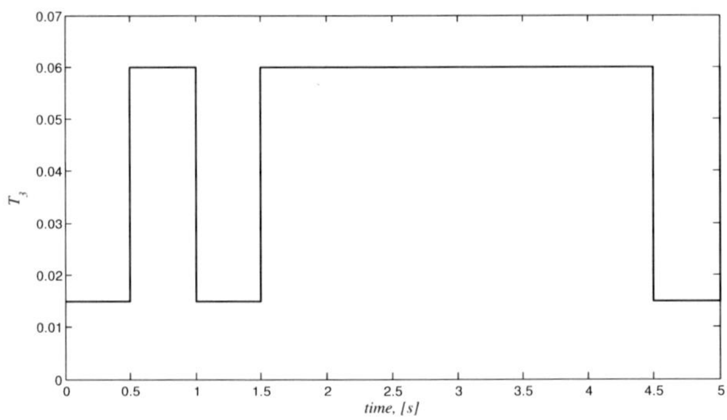

Figure 13. Plant 3 sampling interval over time, with $T_{3,1} = 15$ms and $T_{3,2} = 60$ms.

Conclusion

We proposed a switching output feedback compensator design method for a class of Networked Control Systems. The proposed approach takes into account both induced delays and arbitrary sampling intervals variations over a given set, guaranteeing stability and a certain performance level, aiming at the regulation of network resources. The proposed stabilization result is based on switching parameter dependent Lyapunov functions and is described in terms of a set of LMIs. A switching controller synthesis is then promptly derived leading to delay-dependent compensators. The feasibility of the proposed LMIs certifies the effectiveness of the design method and, furthermore simulations of the NCS can be performed to confirm the expected closed-loop stability under different conditions.

The proposed approach was exemplified by considering a scenario with multiple control tasks, that may compete for the time-varying bandwidth. Such a scenario shows the potential for using our proposal to integrate the control design of various control systems, while also considering the available shared resource in the network. To this end, the presence of a Resource Manager can be considered, as some of its characteristics, such as the interval time that it takes to modify the sampling intervals of the plants, and integrated with the synthesis procedure.

References

Antunes, A., P. Pedreiras, L. Almeida, and A. Mota (2007). Dynamic Rate and Control Adaptation in Networked Control Systems. In *Industrial Informatics, 2007 5th IEEE International Conference on*, Volume 2, pp. 841–846.

Bouyssounouse, B. and J. Sifakis (2005). *Embedded Systems Design: The ARTIST Roadmap for Research and Development*. Lecture Notes in Computer Science / Programming and Software Engineering. Springer.

Boyd, S., L. El Ghaoui, E. Feron, and V. Balakrishnan (1994). *Linear Matrix Inequalities in System and Control Theory*. Philadelphia, USA: Society for Industrial and Applied Mathematics.

Castelan, E. B., V. J. S. Leite, M. F. Miranda, and V. M. Moraes (2010). Synthesis of Output Feedback Controllers for a Class of Nonlinear Parameter-Varying Discrete-Time Systems Subject to Actuators Limitations. In *American Control Conference (ACC), 2010*, Baltimore, USA, pp. 4235–4240. IEEE.

Cervin, A., D. Henriksson, B. Lincoln, J. E. Eker, and K.-E. Årzén (2003). How Does Control Timing Affect Performance? Analysis and Simulation of Timing Using Jitterbug and TrueTime. *IEEE Control Systems Magazine 23*(3), 16–30.

Donkers, M., L. Hetel, W. Heemels, N. van de Wouw, and M. Steinbuch (2009). Stability Analysis of Networked Control Systems Using a Switched Linear

Systems Approach. In R. Majumdar and P. Tabuada (Eds.), *Hybrid Systems: Computation and Control*, Volume 5469 of *Lecture Notes in Computer Science*, pp. 150–164. Springer Berlin / Heidelberg.

Donkers, M. C. F., J. Daafouz, and W. P. M. H. Heemels (2014). Output-Based Controller Synthesis for Networked Control Systems with Periodic Protocols and Time-Varying Transmission Intervals and Delays. *IFAC Proceedings Volumes 47*(3), 6478–6483.

Dritsas, L. and A. Tzes (2007). Robust Output Feedback Control of Networked Systems. In *Proceedings of the 9th European Control Conference*, Kos, Greece, pp. 3939–3945.

Gao, H., X. Meng, T. Chen, and J. Lam (2010). Stabilization of Networked Control Systems via Dynamic Output-Feedback Controllers. *SIAM Journal on Control and Optimization 48*(5), 3643–3658.

Heemels, W. and N. van de Wouw (2010). Stability and Stabilization of Networked Control Systems. In A. Bemporad, M. Heemels, and M. Johansson (Eds.), *Networked Control Systems*, Volume 406 of *Lecture Notes in Control and Information Sciences*, pp. 203–253. Springer Berlin / Heidelberg.

Hetel, L., J. Daafouz, and C. Iung (2006). Stabilization of Arbitrary Switched Linear Systems With Unknown Time-Varying Delays. *Automatic Control, IEEE Transactions on 51*(10), 1668–1674.

Jiang, C., D. Zou, Q. Zhang, and S. Guo (2010). Quantized dynamic output feedback control for networked control systems. *Systems Engineering and Electronics, Journal of 21*(6), 1025–1032.

Jungers, M., E. B. Castelan, V. M. Moraes, and U. F. Moreno (2013). A dynamic output feedback controller for NCS based on delay estimates. *Automatica 49*(3), 788–792.

Mahmoud, M. S. and G. D. Khan (2015). Dynamic output feedback of networked control systems with partially known markov chain packet dropouts. *Optimal Control Applications and Methods 36*(1), 29–44.

Montestruque, L. A. and P. J. Antsaklis (2002). State and output feedback control in model-based networked control systems. In *Decision and Control,*

2002, Proceedings of the 41st IEEE Conference on, Volume 2, Las Vegas, USA, pp. 1620–1625.

Moraes, V. M., E. B. Castelan, and U. F. Moreno (2013). Full-Order Dynamic Output-Feedback Compensator for Time-Stamped Networked Control Systems. *Journal of Control, Automation and Electrical Systems 24*(1-2), 22–32.

Moyne, J. R. and D. M. Tilbury (2007). The Emergence of Industrial Control Networks for Manufacturing Control, Diagnostics, and Safety Data. *Proceedings of the IEEE 95*(1), 29–47.

Naghshtabrizi, P. and J. P. Hespanha (2005). Designing an observer-based controller for a network control system. In *Decision and Control, 2005 and 2005 European Control Conference. CDC-ECC'05. 44th IEEE Conference on*, pp. 848–853.

Rasool, F., D. Huang, and S. K. Nguang (2012). Robust H_∞ output feedback control of networked control systems with multiple quantizers. *Journal of the Franklin Institute 349*(3), 1153–1173.

Scherer, C., P. Gahinet, and M. Chilali (1997). Multiobjective output-feedback control via LMI optimization. *IEEE Transactions on Automatic Control 42*(7), 896–911.

Shi, Y. and B. Yu (2011). Robust mixed H_2/H_∞ control of networked control systems with random time delays in both forward and backward communication links. *Automatica 47*(4), 754–760.

Tang, X. and J. Yu (2007). Networked Control System: Survey and Directions. In K. Li, M. Fei, G. Irwin, and S. Ma (Eds.), *Bio-Inspired Computational Intelligence and Applications*, Volume 4688 of *Lecture Notes in Computer Science*, pp. 473–481. Springer Berlin / Heidelberg.

Velasco, M., J. M. Fuertes, C. Lin, P. Marti, and S. Brandt (2004). A control approach to bandwidth management in networked control systems. In *Industrial Electronics Society, 2004. IECON 2004. 30th Annual Conference of IEEE*, Volume 3, pp. 2343–2348.

Weihua, D. and F. Minrui (2009). Dynamic Output Feedback Control of Uncertain Networked Control Systems. In *Proceedings of the First ACM/SIGEVO*

Summit on Genetic and Evolutionary Computation, New York, USA, pp. 571–576. ACM.

Wittenmark, B., K. J. Åström, and K.-E. Årzén (2002). Computer Control: An Overview. *Technical report*, IFAC Professional Brief, Lund, Sweden.

Yang, T. C. (2006). Networked control system: a brief survey. *IEE Proceedings-Control Theory and Applications 153*(4), 403–412.

Zhang, J., J. Lam, and Y. Xia (2014). Output feedback delay compensation control for networked control systems with random delays. *Information Sciences 265*(0), 154–166.

Zhang, L., H. Gao, and O. Kaynak (2013). Network-Induced Constraints in Networked Control Systems: A Survey. *Industrial Informatics, IEEE Transactions on 9*(1), 403–416.

Zhang, X. M., Q. L. Han, and X. Yu (2016). Survey on recent advances in networked control systems. *IEEE Transactions on Industrial Informatics 12*(5), 1740–1752.

In: Networked Control Systems
Editor: Eduardo Paciência Godoy
ISBN: 978-1-53613-105-5
© 2018 Nova Science Publishers, Inc.

Chapter 3

Wireless Networked Control Systems: Principles and Evaluation

***Guilherme Bertelli*[1,*] *and Ivanovitch Silva*[2]**

[1]Nucleus of Research and Innovation in Information Technology,
Graduate Program in Electrical and Computer Engineering,
Federal University of Rio Grande do Norte, Natal, RN, Brazil
[2]Digital Metropolis Institute,
Federal University of Rio Grande do Norte, Natal, RN, Brazil

Abstract

With the rise of Industry 4.0, wireless communication is a tendency in industrial environments, saving resources and enabling the addition of new applications when compared to their wired counterparts. In this context, the WirelessHART and ISA100.11a specifications are emerging as solutions for the last mile connection. Despite their high degree of applicability, due to the nature of its technical challenges (reliability, energy consumption, fault tolerance, ensuring constant delays), industrial wireless networks used, in the majority of applications, for monitoring processes. Given the aforementioned and the conservatory tendency of the

* Corresponding Email: guilherme@dca.ufrn.br.

industry, the demand for evaluation and assured reliability tools on Wireless Networked Control Systems (WNCS) is imminent. Aiming to enlighten and demystify the use of wireless technology on the control loops in the industry, this work presents an overview of WNCS and a practical experiment. The analysis of the behavior of typical WirelessHART network in process control is done by controlling the level loop of a process station through a PID algorithm implemented in a host application. The conducted control study used the Dust-Networks SmartMesh WirelessHART Starter Kit to establish the network, allied to Emerson's 3051S Pressure Transmitter, and assures reliability and efficiency for WNCS applications.

Keywords: Wireless Networked Control Systems, Industrial Wireless Technologies, WirelessHART, ISA100.11a.

1. Introduction

With the concepts of Internet of Things and cyber-physical systems being introduced to industrial scenarios, industrial automation is going through a big revolution, called the Industry 4.0 [1]. In this context, wireless revolution has transformed everyday activities, connecting people full-time, but has not yet made large inroads into manufacturing. The maturation of wireless networks technology, along the incoming years, has been able to successfully, albeit slowly, demonstrate efficiency in the demanding process plant environment. Wireless communication to and from the final control element is now commercially available, leading to an increase interest and study in Wireless Networked Control Systems (WNCS), although there is not yet a universally accepted standard for wireless control.

Albeit eliminating 50-90% of installation and maintenance costs [8], the use of wireless technologies in industrial environments has always been viewed with great skepticism by plant managers. These concerns were created primarily by the unreliable, non-deterministic nature of the communication channel [3]. The situation is aggravated by the fact that the equipment are installed in areas subject to the influence of external agents (noise, interference, adverse weather, natural obstacles), which can generate high transmission error rates when compared to cabled technologies [4]. Other communication errors might occur due to signal attenuation (influenced by the distance between the transmitter and receptor) and the problem of multi-path fading (due to reflection, diffraction and spreading of the transmitted signal) [5].

Other relevant issues to be analyzed are dependability requirements (reliability and availability), as faults may lead to system failures, which can result in financial losses, environmental damage, or putting people in danger [9]. Thus, a person who is familiar with traditional control techniques may be reluctant to take the time needed to explore how wireless measurement and control elements may be used in closed loop control.

Because of the conservative nature of the process industry, wireless field devices were initially installed to better monitor process operation. However, as plants gained experience with wireless technology, there has been a growing interest in using wireless sensors and actuators for process control. The necessity for extensibility, mobility, modularity, fast deployment, and reduced installation and maintenance cost are key factors behind this change in attitude [6].

In spite of all those challenges, the evolution of wireless technology has led to the development of new mechanisms to ensure the reliability of the network, such as modulation, coexistence with other technologies, cryptography, deterministic schedule, frequency hops and redundancy [11, 12]. Those changes allied to the creation of international standards, such as WirelessHART and ISA100.11a, made the use of wireless technology viable in industrial environments.

Given the necessity of the industry to use wireless technologies for WNCS, this chapter presents an overview of wireless networked control systems and a real-time evaluation of a feedback level control system over a WirelessHART network. The goal is to analyze the system's performance, transport delays and the reliability of industrial wireless networks for control applications.

The paper is organized as followed: section 2 presents an overview of wireless networked control and its technologies. Section 3 presents the related works in industrial wireless networks evaluation for process control.

The WirelessHART evaluation testbed is described in section 4. The analysis and results are discussed in section 5. Finally, section 6 presents a conclusion, contributions and future works.

2. Wireless Networked Control Systems

A Networked Control System (NCS) is characterized by the insertion of a communication network in the feedback control loop, which handles the information transfer [16]. In a WNCS, the communication is done via a wireless channel.

In NCS, there are two patterns of communications: the direct structure and hierarchical structure [7]. The direct structure shares the network resources with sensors, actuators and field devices, while in the hierarchical structure a main controller coordinates the operations of local controllers over a network.

Given the challenges and criticality of control applications, it is important to understand the possible architectures and recent research approaches to be taken in consideration for WNCS design.

2.1. Network Architecture and Design Approaches

The network architecture is an important topic of study for the performance of NCS and its integration with legacy systems. Traditionally, the control algorithm is implemented in a dedicated controller, though in a WNCS more options are available. Figure 1 presents typical WNCS formats.

The control algorithm can be allocated in the gateway, with the network architecture following a direct structure, as seen in Figure 1(a). The controller can be implemented in a field device (mote, embedded system, etc.) either in the sensor or in the actuator, illustrated in Figure 1(b) and 1(c), respectively. In addition, the controller can be implemented in the plant automation network, shown in Figure 1(d). This case is found in conventional NCS schemes, with measuring and actuation via cache memory updated periodically in the gateway.

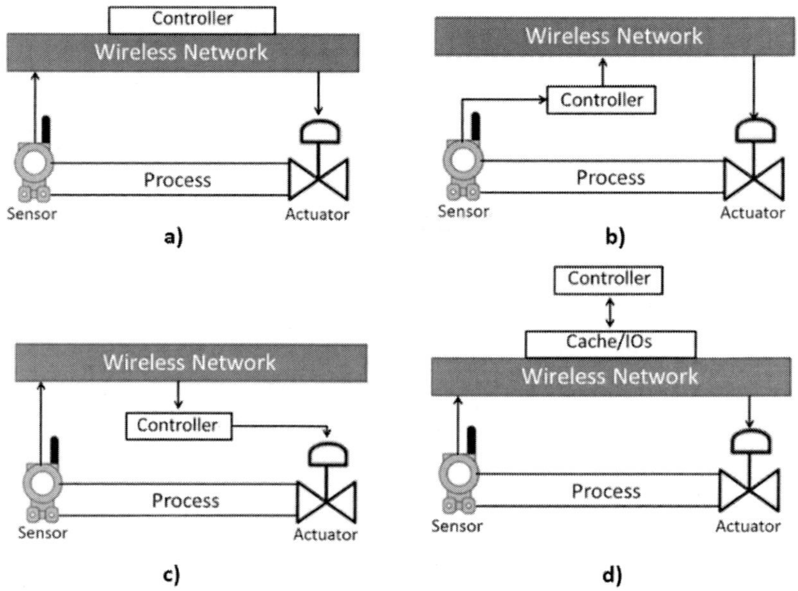

Figure 1. WNCS Architectures. (a) Controller implemented in the Gateway; (b, c) Controller implemented in field device; (d) Controller implemented in the automation network.

The control algorithm can be allocated in the gateway, with the network architecture following a direct structure, as seen in Figure 1(a). The controller can be implemented in a field device (mote, embedded system, etc.) either in the sensor or in the actuator, illustrated in Figure 1(b) and 1(c), respectively. In addition, the controller can be implemented in the plant automation network, shown in Figure 1(d). This case is found in conventional NCS schemes, with measuring and actuation via cache memory updated periodically in the gateway.

The non-deterministic nature of a shared wireless communication network does not assure instant communication service without data loss, and are subject to delays [18]. Considering those concerns, three NCS design approaches can be found in literature [7, 21]:

- Network Aware Control: this approach focuses on design/selection of communication protocols and network infrastructure. Network topology, real-time scheduling, control structure and architecture issues should be studied [26].
- Control Aware Network: this strategy compensates channel issues with the controller design. Time delays, data loss probability and other non-deterministic challenges are modeled [27, 32]; sampling, control implementation and actuations are linked to scheduling.
- Co-Design Approach: a more recent approach, involves study and design of both network and controller to address shortcomings that the other approaches have with control-over-wireless issues [36]. The strategy involves the design of a failure-tolerant controller and an optimized network for control performance, providing an adaptive control layer [38].

The use of wireless networks for monitoring is currently largely applied without major concerns, but its application in process control is challenging. Due to the real-time and reliability requirements of process control, the challenges of a wireless network are more likely to derail a wireless system then they would in a traditional wired channel [17].

2.2. Challenges

In a shared network, the data packets are subjected to network induced delays, processing delays, and may even experience data-loss, which can degrade the system performance and lead to instability [18]. In general, errors in wireless communications are transient; that is, the communication channel is bad for a time and then returns to normal. In an opposite way, communication errors in wired channels are usually permanent due to faults in cables, connectors, or other components [9].

Security is always an issue to consider when using wireless communications. As the communication signal is transmitted through the air, any device within range can receive the signal and eventually decode it.

This could lead to problems of unauthorized access to sensitive information. In the end, unauthorized persons can take advantage of the absence of security measures to inject packets into the network in order to perform attacks or to steal information. Thus, it is essential to consider the use of security measures to avoid these problems [10].

Energy consumption is also a priority, as wireless networks devices are battery-based. Therefore, control algorithm and network operation must be energy-efficient, minimizing communication while maintaining control performance [21].

Another challenge is related to the coexistence of different wireless technologies in the same area. Since the transmission medium is open, there may be situations where different technologies share the same frequency range. Thus, it is important that even when they coexist in the same environment, different technologies can operate without interferences [9].

New design techniques have been developed for a more efficient control in a wireless network, such as PIDPlus and Model Predictive Wireless Control [2], though a standard PID controller may be used if the systems response is at least four times slower than the communication update rate [19]. An efficient wireless control has requirements on the following specifications, in general terms [20]:

- Rate and latency: sampling rates of 1-2 seconds, controlled latency (~50% of sampling rate), 4 Hz sampling in constrained configurations;
- Flexible system architecture: engineered and scalable IP backbone;
- Mesh networking: interoperable peer-to-peer connections, function blocks at the device level, deterministic battery life;
- Reliability: deterministic and accurate wireless transmission;
- Security: encrypted/anti-hacking wireless transmission.

The challenges related to the demand for using wireless technology led to the development of wireless communication standards and the two main technologies used in industrial environments are WirelessHART and ISA.100.11.a. These wireless protocols are based on IEEE 802.15.4e standards, and institutions such as IEEE and IETF have been working to adapt the traditional IEEE 802.15.4 standard to non-critical industrial scenarios.

2.3. Industrial Wireless Technologies

2.3.1. WirelessHART

WirelessHART is an industrial wireless communication standard developed by the HART Communication Foundation (HCF) with the goal of transmitting HART messages without the legacy transmission means (4-20 mA or RS484). A WirelessHART device implements the same command structure as a HART device, therefore being limited to the applications supported by a traditional HART network. In September 2008, the WirelessHART standard (HART 7.1) was approved by the International Electrotechnical Commission (IEC) as a publicly available specification (IEC 62591), becoming the first industrial wireless communication technology to attain this level of international recognition. The final version of the WirelessHART standard was approved at the beginning of 2010 [13].

WirelessHART defines eight types of devices: network manager, network security, gateway, access point, field device, adapter, router, and handheld device. All devices that are connected to the wireless network implement basic mechanisms to support network formation, maintenance, routing, security, and reliability. A typical WirelessHART network can be seen in Figure 2.

Field devices are the most basic WirelessHART devices. They are directly connected to the process and plant equipment. Field devices can transmit measurement data and receive and forward packets to and from any device. Usually they can be line, loop, or battery powered. Compatibility with legacy HART devices is guaranteed with adapter devices. The adapter devices are not directly connected to the plant equipment; however, they have to support the same functionalities as field devices. On the other hand, handheld devices are used during the network's installation, configuration, and maintenance phases. They do not support routing mechanisms.

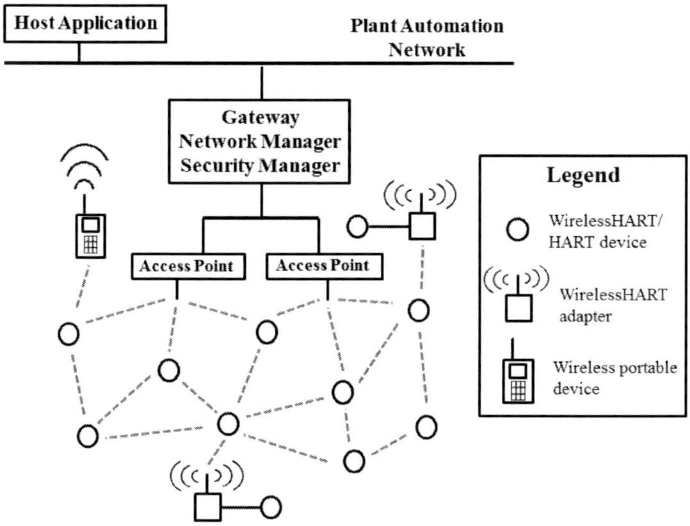

Figure 2. Typical WirelessHART network.

Router devices are used for routing purposes, that is, forwarding packets from one device to another. They are not directly connected to the industrial process, and thus they can be installed anywhere in the plant. Their use is not necessary since field devices have internal routing capabilities. However, router devices can provide redundant paths to the gateway, and they can minimize energy consumption in field devices. The gateway works as a sink point for all wireless traffic and provides the connection between the plant automation network and the wireless network. The logical communication between the gateway and the wireless network occurs through access points installed in the gateway. The amount of access points can be configured to increase redundancy and to improve the effective network throughput.

The security manager is the entity responsible for ensuring the security over the network. It provides services such as joining and network and session keys for all devices. These keys are used to authenticate and encrypt data. The storage and management of keys are also under the responsibility of the security manager.

The core of the WirelessHART is the network manager. It is logically connected to the gateway and manages the entire network. The communication between the network manager and the network devices occurs through the application layer protocol. The main duties of the network manager are related to scheduling, management of the device list, routing (redundant paths), and collecting information about performance, failure detection, and network formation.

2.3.2. ISA100.11a

Approved by the International Society of Automation (ISA), the ISA100.11a is currently named in the draft version as IEC/PAS 62734 [14], after the International Electrotechnical Commission (IEC) created a task force to approve ISA100.11a as a publicly available specification in 2011. Differently from the WirelessHART standard, whose development was based on a specific technology (HART), the ISA100.11a standard was designed based on the user application requirements [9].

The main goals of the ISA100.11a standard are related to guaranteeing the deterministic communication latency while increasing the communication reliability, focusing on process control and monitoring applications, with latency requirements of around 100ms. ISA100.11a can coexist with other wireless technologies such as cell phones, IEEE 802.11, IEEE 802.15, and IEEE 802.16, and provides tunneling for legacy protocols (HART, Foundation Fieldbus, Profibus, and Modbus).

A typical ISA 100.11a network, which is described in Figure 3, may be composed of seven types of devices: gateway, system manager, security manager, router, backbone router, Input/output (IO) devices, and portable devices. Each device has a specific role definition that controls its functions. The I/O device is responsible for monitoring the environment and it has no routing function to minimize the energy consumption. It can also be portable, and in this case, its operation is limited to testing, management, and configuration of other devices.

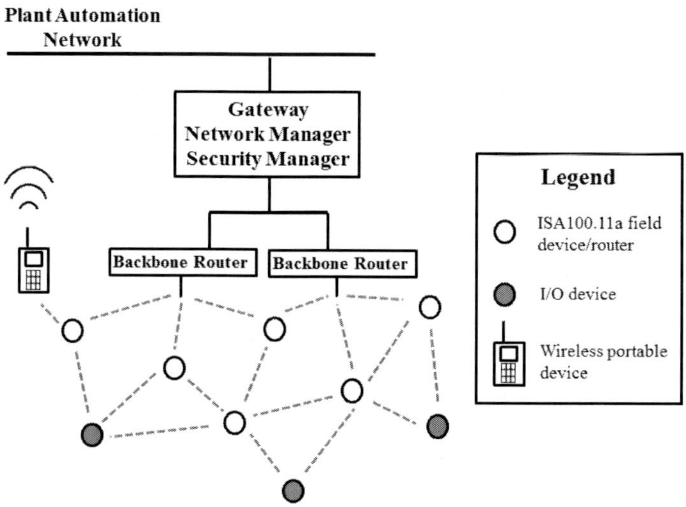

Figure 3. Typical ISA100.11a network.

Two devices conduct the routing mechanism: the router and the backbone router. The former works as a proxy, forwarding the network data towards the gateway, whereas the latter is responsible for high level routing between the gateway and the backbone routers. All traffic between the backbone routers is based on 6LoWPAN (IPV6 over low power wireless personal area networks) [15]. Additionally, the backbone routers can segregate the network into sub-nets. Similarly, to the WirelessHART network, an ISA100.11a network is centralized to three devices: the gateway, security manager, and system manager.

The gateway device provides a connection between the wireless sensor network and the plant automation network. It is also responsible for guaranteeing the tunneling with the legacy technologies. Due to its importance, there is support for redundant gateways. The security manager and the system manager perform the most important tasks. The security manager controls the system security management function, whereas the system manager governs all the network, devices, and communications.

Although the ISA.100.11.a technology has proven to be reliable, the control systems in process industry can be quite complex. Therefore, given how familiar traditional control techniques are, companies tend to be very reluctant to invest the time needed to apply wireless measurement and control elements to closed-loop control.

2.3.3. WirelessHART and ISA100.11a Comparison

The wireless sensor networks were designed to support devices with limited physical capabilities and resilient communications approaches. These networks enable the development of emerging industrial applications where the use of legacy wired technologies is difficult or impracticable, for example in tank farms, pipelines, distillation columns, wellheads, and rotation equipment (turbines and kilns). In this sense, all of the technologies described in the previous sections can be adopted as communication solutions for these emerging industrial applications. Thus, it is important to compare the main features of these technologies to aid the design of applications. Considering the protocol stack, the more relevant information about the WirelessHART and ISA100.11a technologies is summarized in Table 1 [9].

Table 1. Wireless standards comparison

	Industrial Wireless Standards	
Layer	WirelessHART	ISA100.11a
Applications	Command-oriented	Object-oriented
Transport	End-to-end reliability	UDP extension
Network	Mesh (graph and source routing)	High-level routing (backbone routers)
Data link	TDMA, CSMA-CA, 10 ms time slot, frequency hopping (1 profile)	TDMA, CSMA-CA, 10 or 12 ms time slot, frequency hopping (3 profiles), mesh (graph and source routing)
Physical	IEEE 802.15.4	IEEE 802.15.4

With both standards supporting a physical layer based on IEEE 802.15.4, their upper layers have similar approaches. The WirelessHART and ISA100.11a networks implement the same mesh routing algorithm, with the latter also supporting subnets. Additionally, the ISA100.11a implements a frequency-hopping scheme more flexible than the WirelessHART standard. It is worth noting that only in the WirelessHART standard it is possible for a field device to accumulate the routing function.

Regarding the application layer, the WirelessHART standard was designed to perform only HART applications. Thus, its application layer is limited to commands supported by a HART network. On the other hand, a flexible object-oriented approach was developed in the application layers of ISA100.11a and WIA-PA standards. Thus, it is possible to use legacy applications (HART, Modbus, Foundation Fieldbus, and Profibus) over ISA100.11a networks. With the given context, the next section will bring an overview of the latest works on wireless networked control systems performance evaluation.

3. Related Works

The research directions that are identified related to the design and implementation of wireless process automation networks to enhance control-over-wireless are [21]:

- Energy-efficient control algorithms to minimize communications across the network;
- Design of latency control mechanism;
- Mathematical formulation of theoretical limit on the size of a network to meet a latency specification;
- Closed-loop control algorithms when delays are uncertain or unbounded.

This section will present the latest research approaches in the area of WNCS performance evaluation, divided into two main topics: wireless network evaluation and control over wireless networks analysis.

3.1. Wireless Networks Evaluation

In the paper [43], a new module for the NS-3 simulator for WirelessHART network evaluation was proposed. The module permits configuring each link with different fault probabilities, allowing simulations that are more accurate. For validation, the scheduling and routing approaches were statically configured assuming different topologies and probabilistic error scenarios.

The authors in [45] evaluated an ISA100.11a network MAC and physical layer based on accurate simulation model using OPNET modeler. Several sets of simulations with different network settings are conducted to study its various features, including the effects of timeslot duration, superframe period, and maximum back off exponent. The network throughput, average delay, and energy consumption/superframe are investigated as the performance metrics.

A software to obtain network data from a WirelessHART network is presented in [39]. The developed software allows identifying relevant issues to the verification and maintenance of WirelessHART networks. This research group also developed a monitoring software architecture for inspection of WirelessHART networks [40], in which the capture of information is done in a passive way using sniffers deployed in the network area.

In order to satisfy real-time communication requirements, the work [41] proposes efficient routing algorithms, defining uplink, downlink and broadcast graphs to generate data link layer communication and achieve end-to-end real-time performance.

Petersen et al. did an interesting and important study on the reliability of WirelessHART networks in [35]. The study tested a WirelessHART network with 9 nodes and one Gateway coexisting with the IEEE 802.11 standard. Interference tests were also applied by chirp jamming.

In [46], a methodology based on an automatic generation of a fault tree to evaluate the reliability and availability of Wireless Sensor Networks (WSN) was developed. This proposal supports any topology, different levels of redundancy, network reconfigurations, criticality of devices and arbitrary failure conditions of WirelessHART and ISA100.11a networks.

Different strategies can be associated to WirelessHART networks routing and scheduling algorithms, and the most relevant works were summarized in [37]. This study confirms how broad this field of research is, with applications ranging from spatial redundancy for monitoring applications to temporal redundancy for control applications.

3.2 Control Over Wireless Networks Analysis

Given the importance of process control through industrial networks and the increasing use of wireless networks in the industry, the necessity for an evaluation of communication delays and reliability of those system for control applications in imminent. It is desirable to analyze the performance in several configurations (different time intervals) to prevent failures and enhance the performance according to the analyzed state.

In a broader analysis, networked control in general has been approached in a vast number of studies, which are summarized in the work [21]. This thesis provides a deep overview of all the background on the main topics in networked control while addressing to three issues: time synchronization, study of the performance of closed-loop applications running over a resource-constrained wireless network and ensure longevity of wireless field nodes operated on standalone batteries, which requires mechanism to minimize energy consumption. The thesis also offers an insight about the future directions of wireless communications in the process industries.

This necessity has led to the development of projects with simulation of WirelessHART networks applied to control systems [22, 23], extending the Simulink True Time tool from the software MATLAB, with tests being performed with a fixed number of lost packets. In [22] specifically, the author also conducted a comparison between the performance of WirelessHART and ZigBee networks.

An analysis of controllers using WirelessHART networks was conducted in [24]. The authors developed a simulation model that sets up

WirelessHART networks in different configurations. The communication channels were assessed according to the size of the network and location of each device in order to determine the best scaling method for control applications. This proposal also draws firm conclusions about wireless networked control.

Another work using TrueTime for simulation control systems in WirelessHART networks was proposed in [25]. In that paper, the features of the simulator were extended by developing a tool to introduce more detail to the simulated environment and evaluate problems regarding the coexistence of wireless networks in the same area.

The paper [33] refers to Optimal Controller Location (OCL) in WNCS, presenting two approaches based on linear-quadratic-Gaussian applicable to One-hop and Multi-hop networks. The results turn the OCL problem into a finite-solution optimization problem that can be solved efficiently.

A performance analysis on WNCS was conducted in the thesis [34], presenting new strategies for packet transmission in multi-hop wireless networks. By handling experiments with classic and event-triggered controllers, the authors presents solid conclusions on the use of WirelessHART and ISA100.11a standards.

Blevins et al. has done an extensive study in his book [2], with technical evaluation of WirelessHART network in control loops, going through many case scenarios and wired vs. wired comparisons.

The authors in [16] proposed a WirelessHART networked control through a host application in a practical experiment. The application was used to control the position of a proportional valve, with a solid analysis of the controller's sample time and the system's delays. Another WirelessHART control-loop evaluation through host application was conducted in [44], with the goal of demystifying the use of wireless technologies for control applications. Different sampling times were tested, showing an efficient control, albeit with sampling limitations.

A different WirelessHART control-loop evaluation was accomplished in [42] using a *Tmote Sky* to implement the network. A LQR controller with a Kalman filter was implemented, prioritizing temporal redundancy and latency time.

Table 2. Related works summary

Wireless Network	Study	Author(s)
WirelessHART	WirelessHART network evaluation software	[39, 40, 43]
	WirelessHART network coexistence evaluation with the IEEE 802.11 standard	[35]
	Survey on routing and scheduling algorithms	[37]
	Simulation of WirelessHART networks applied to control systems	[22, 23], [24, 25]
	Technical evaluation of WirelessHART network in control loops	[2]
	Performance evaluation on control-loop through host application	[16, 44]
	WirelessHART network implementation with *Tmote sky* + LQR controller	[42]
ISA100.11a	ISA100.11a network MAC and physical layer evaluation based on simulation model	[45]
	Performance evaluation on control-loop through host application	[28]
	ISA100.11a latency evaluation for control applications	[6]
WirelessHART and ISA100.11a	Study of time synchronization, performance of closed-loop and energy consumption	[21]
Wireless Network	Study	Author(s)
WirelessHART and ISA100.11a	New strategies for packet transmission in multi-hop wireless networks	[34]
	Efficient routing algorithms to achieve end-to-end real-time performance	[41]
	Approaches on OCL in WNCS applicable to One-hop and Multi-hop networks	[33]
	Automatic generation of a fault tree to evaluate the reliability and availability of WSN	[46]

The ISA100.11a standard was addressed in [6], where the authors evaluate the suitability of ISA100.11a network for use in process control through real-time experiments, analyzing coexistence with WirelessHART and Wi-Fi nodes and latency requirements. In a previous work, the authors conducted an ISA100.11a performance evaluation in [28]. The goal was to analyze the system's dynamic response with different controller sampling times, with the network established with real industrial instruments and devices. The results proved the reliability and efficiency of the ISA100.11a standard for control applications, though with sampling time constraints.

Table 2 presents a synopsis of the developed research in performance evaluation of WirelessHART and ISA100.11a technologies.

4. WirelessHART Testbed

Aiming to demystify the use of wireless technology on industrial process control, a real-time evaluation of feedback level control through a WirelessHART network was conducted, with the goal of analyzing the system's performance, transport delays and the reliability of industrial wireless networks for control applications.

The supervisory application, which includes the control action algorithm (discrete PID control), was developed by the author's research group, in Java. The controller interface is shown in Figure 4.

The proposal was validated with a SmartMesh WirelessHART Starter kit by Dust-Network, a BeagleBone Black, and an Emerson 3051S Pressure Transmitter. In order to assure reliability and efficiency in industrial wireless networks for control applications, the most common and well-known controller was implemented: a traditional PID controller with anti-windup filter.

The developed control application communicates with the WirelessHART network from a universal communication driver based on XML-RPC. The control tests were conducted in a MPS-PA Process Station [30] by Festo. The Beaglebone Black was used to access the data from the motes via UART communication. The scenario is shown in Figure 5, which is equivalent to the architecture depicted in Figure 1d), where the controller is implemented in the automation network.

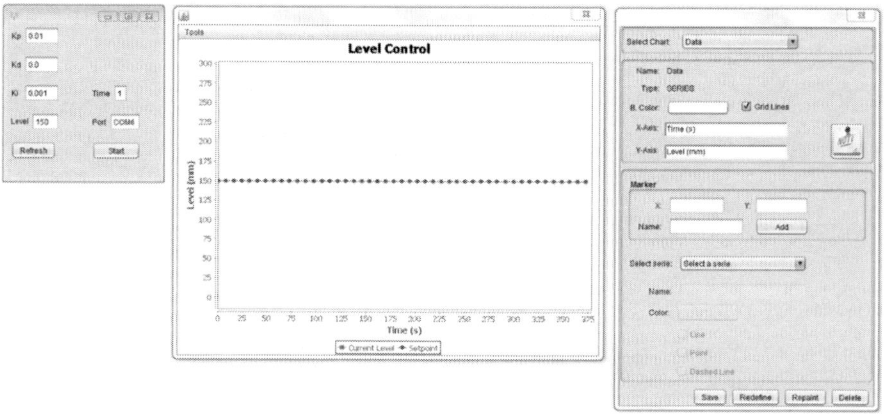

Figure 4. Controller interface.

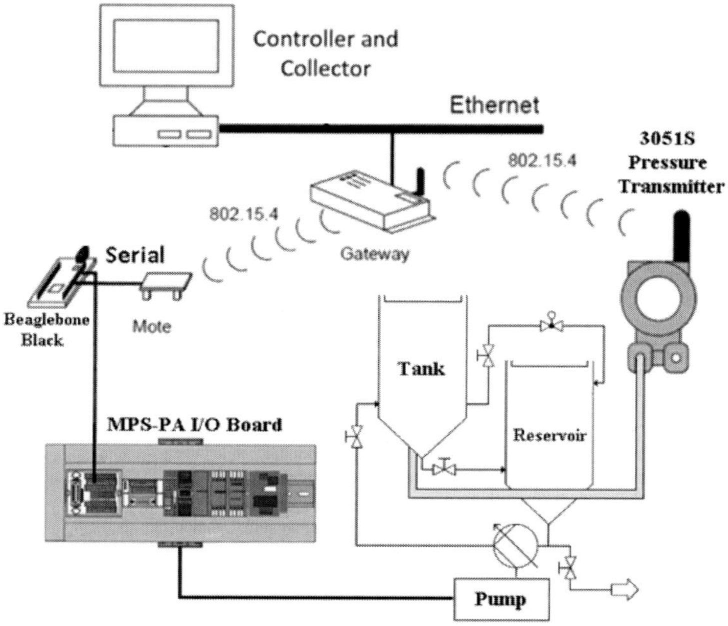

Figure 5. WirelessHART experimental scenario.

4.1. Dust-Networks SmartMesh WirelessHART Starter Kit

The Dust-Linear SmartMesh WirelessHART Starter Kit [31] provides the necessary devices to build a typical WirelessHART network The kit was designed for the harshest industrial environments, where low-power, reliability, resilience and scalability are key.

The kit (seen in Figure 6) includes a manager (LTP5903CEN), providing core networking functionality, enabling the network to achieve unsurpassed levels of resilience, reliability and scalability. The kit also includes an ultra-low power mote (DC9003A-C) which are the field devices that populate the network and an interface card (Eterna DC9006), which provides a USB connection for both power and software access to the mote API UART port.

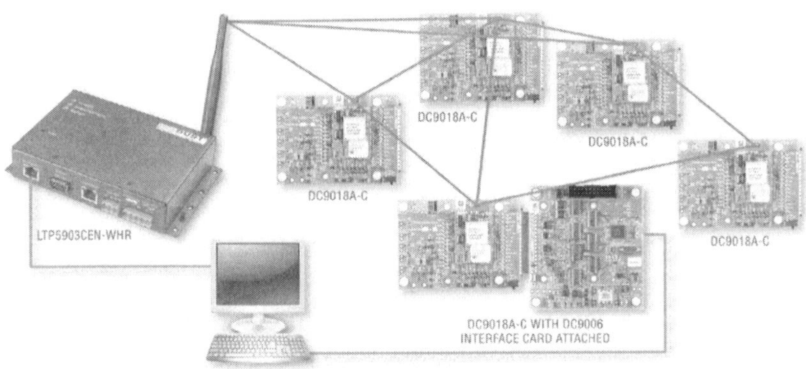

Figure 6. Dust-Networks SmartMesh WirelessHART Starter Kit [31].

4.2. Device Communication and XML-RPC

The developed control application communicates with the WirelessHART network from a universal communication driver based on XML-RPC. This driver is based on the XML-RPC protocol, consisting of remote procedure calls (RPC) encoded in XML over HTTP. This technology uses the combination of three patterns: the communication architecture (RPC), language (XML), and the transport mechanism (HTTP), often used for establishing various types of connections across different devices [29].

XML-RPC is used to interact with SmartMesh WirelessHART Starter Kit network manager and devices, being the simplest mode of communication available (Serial, SSH, and XML-RPC are the available protocols). The client uses procedures provided by a server in a particular URL via HTTP. In the experiments, the gateway acts as a server, while the driver, which aims to get data allowing further analysis, acts as a client [44].

The Beaglebone Black board is used to gather information from the mote, given its lack of analog output. That is accomplished through a serial (UART) connection. The mote must be set in slave mode, with the *trace* function activated by accessing the mote via serial port and using the *loc trace on* command. The *trace* function enables the mote to track all the sent and received information, writing them in the UART ports.

4.3. WirelessHART Control Loop

The Emerson Process Management 3051S transmitter reads the level in the tank and sends this information to the gateway, which manages the network. The implemented controller communicates with the gateway through a data collector, allowing a connection to be opened with the gateway via XML-RPC, enabling asynchronous reception of data from the gateway.

The control software, on the client's computer, receives and processes pressure information from the gateway and generates the control action, sending it back to the gateway, which sends it to the mote. The Beaglebone Black is used to obtain that information via serial connection (UART) from the mote, then processes the data and sends it via its PWM output to the MPS-PA I/O board.

The process station has two modes: a digital mode, for ON/OFF control, and an analog mode, for PID control. To enable the analog mode, the K1 relay from the I/O board must be activated, which is done by a 24V signal being applied to the digital input *XMAX1-02*. Additionally, a 0-10 V PWM signal must be sent to the *X2-1* analog input, controlling the pump. Due to the Beaglebone's Black output limit of 3.3V, the analogue activation signal and the PWM control signal must be amplified. As in the previous case study, the control application updates the PID value each sampling time and calculates the transport delays of the network based on packet flow.

5. Analysis and Results

The WirelessHART network was established using the tools presented in the previous sections. Initially, the gateway was configured through a serial connection. The transmitter was calibrated and commissioned through a software provided by Yokogawa, and joined the network by setting the *network id* and *join key*. The same procedure was applied to the mote, which was accessed via serial connection and set to *slave* mode, its network parameters were configured and it was able to join the network by manually using the command *loc 0600*.

The tests were executed with two different sampling times: 1 and 4 seconds, with a set point varying each 180 seconds (150 mm, 250 mm, and 100 mm). The PID controller parameters, obtained experimentally, were $K_p = 8$, $K_i = 1.5$ and $K_d = 0$ for both sampling times. For the 1 second sampling time scenario, the *fastpipe()* function, available with the network manager, was used in order for the gateway to handle more data.

Figure 7 shows the systems response with a 4 second sampling time, followed by Figure 8, which illustrates the packet delays of the system. The dynamic response of the process was satisfactory, having converged to the desired set-points with 124s, 127s and 138s settling times. The overshoots were reasonable, at 22.3%, 11.2% and 22.5%, for each setpoint, respectively. A total of 136 control signal packets were transmitted were an average transport delay of 1.63s.

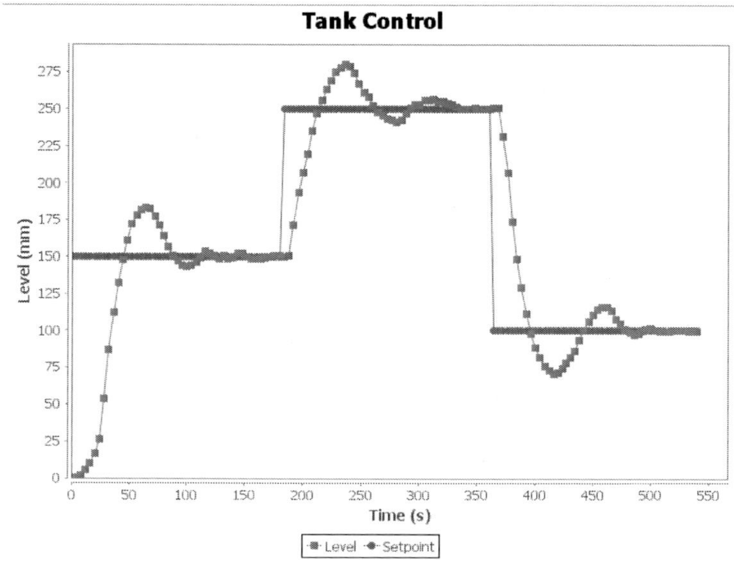

Figure 7. WirelessHART controlled system with 4 seconds sampling time.

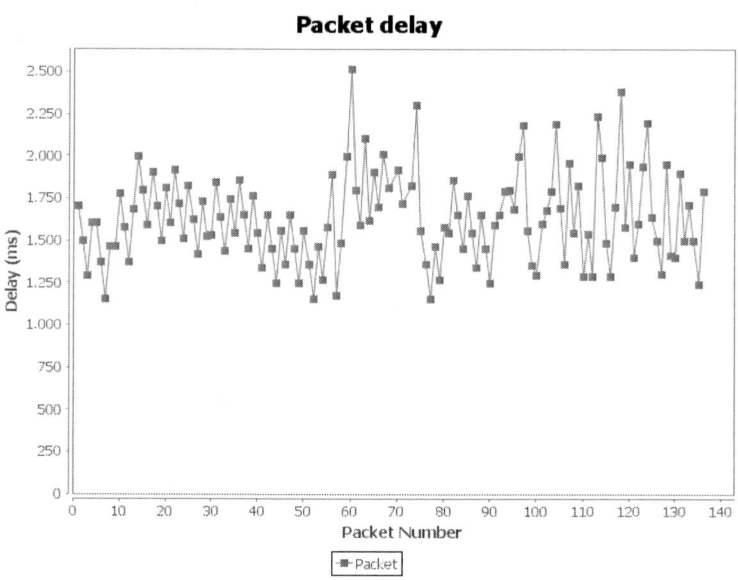

Figure 8. WirelessHART transmission delays with 4 seconds sampling time.

The second test, Figure 10 and Figure 11, was conducted with a sampling time of 1 second, using the Dust-Networks WirelessHART kit function that allows the network manager to have a high-bandwidth communication channel between the gateway and a selected field device, enabling the exchange of a large amount of information at a fast rate. That function is called *fastpipe()*, and can be activated through the manager's console with the command *activateFastpipe("device address")*. The selected device was the mote connected to the Beaglebone Black.

With the *faspipe()* function activated, results were improved. The plant's level was smoothly controlled, as can be seen in Figure 9. The overshoot, for each changing set-point, was 4.5%, 10.3% and 12%. The system settled considerable quicker, with settling times of 63s, 53s and 48s. That can be attributed to the faster communication rate and network packet transmission due to the use of the *fastpipe()* function, drastically reducing transport delays, which had an average time of 292ms, as shown in Figure 10.

It is worth noting, though, that the Emerson Pressure Transmitter 3051S has a measurement sampling time limitation of 1 second, implying that even if the control signal algorithm had a faster sampling rate, it's value would still only update every 1 second.

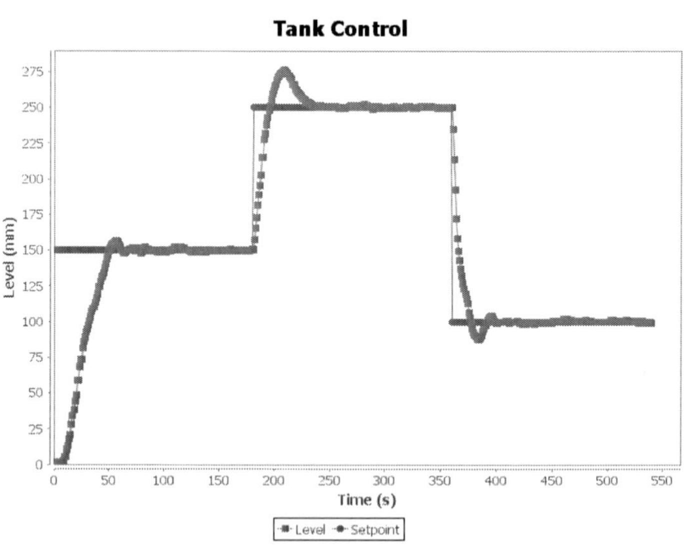

Figure 9. WirelessHART controlled system with 1 second sampling time using *fastpipe()*.

The average performance parameters of the WNCS tests were summarized and can be analyzed in Table 3. The overall performance of the closed-loop networked control was satisfactory.

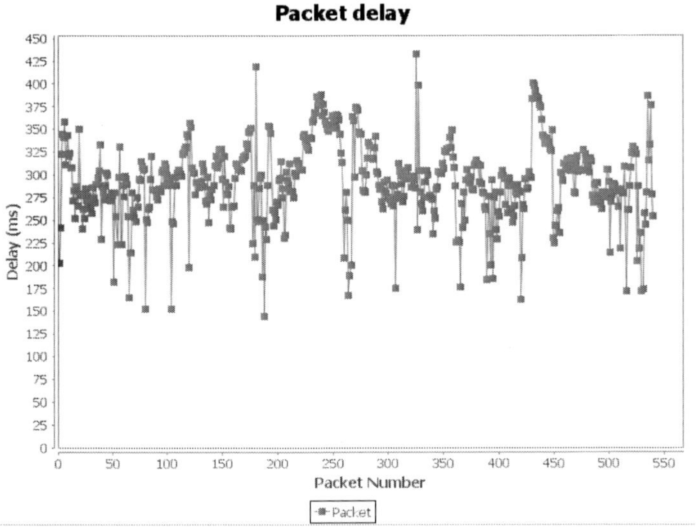

Figure 10. WirelessHART transmission delays with 1 second sampling time using *fastpipe()*.

Table 3. WirelessHART tests summary

	Controller sampling time	
Results	4 seconds	1 second (*fastpipe*)
Overshoot (average)	19.3%	8.9%
Settling time (average)	129s	54s
Delay (average)	1.63s	0.292s

CONCLUSION

With factory automation in context to Industry 4.0, in which connectivity is the tendency, wireless communication is an imminent path for research and development. Through wireless monitoring has been largely accepted, the challenges of control-over-wireless creates skepticism from factory managers albeit creating a less polluted, more reliable and secure environment and reducing costs.

With the establishment of industrial wireless standards such as ISA100.11a and WirelessHART, created for real-time, reliable industrial monitoring and control, the pieces are set for the study and implementation of WNCS. This chapter presents a theoretical overview of WNCS, along with a state-of-the-art on industrial wireless networks evaluation for control applications and a real-time performance evaluation experiment.

The contributions of the results affirm applicability and, more importantly, efficiency and reliability of a WirelessHART network for closed-loop level control by a simple discrete PID algorithm. The plant's dynamic response was adequate for a level system and the latency was limited, with the controller sampling time being limited by the Emerson pressure transmitter.

The challenges of packet loss, communication delays, energy consumption and security are being dealt with research on design of network infrastructure (network aware control approach), modelling of time delays, data loss probability and other non-deterministic challenges for controller design (control aware network approach) and joined study of both network and controller for optimum real-time performance and reliability (co-design approach).

For future works, it is intended to replicate the experiments conducted in [28], which consists an ISA100.11a performance evaluation through host application by a discrete PID controller, but using the same plant as in this work, for a more solid comparison of the WirelessHART and ISA100.11a networks on control-loops. The objective is to analyze how network specifications and architecture might affect a system's dynamic response, which can lead to solutions on controller design and wireless standards updates for control applications. In addition, a more detailed analysis of the latency jitter and the sources of delays is intended.

ACKNOWLEDGMENT

The authors would like to thank Daniel Lopes, Anderson Santos and Heitor Florencio, PhD researchers at LAMP-UFRN, for the help regarding wireless networks configuration and device communication; CAPES, for the financial support through research scholarships; and UFRN for the infrastructure that served as basis for the development of this work.

REFERENCES

[1] Wollschlaeger, M.; Sauter, T.; Jasperneite, J. 2017. "The Future of Industrial Communication: Automation Networks in the Era of Internet of Things and Industry 4.0." *IEEE Industrial Electronics Magazine*, 11(1), 17-27.

[2] Blevins, T. L.; Chen, D.; Nixon, M.; Wojsznis, W. 2015. *Wireless Control Foundation: Continuous and Discrete Control for the Process Industry*. Pittsburgh: International Society of Automation.

[3] Liu, Z.; Ma, M.; Dai, J. 2012. "Utility-based scheduling in wireless multi-hop networks over non-deterministic fading channels." *Computer Networks*, 56(9), 2304-2315.

[4] Bai, H.; Atiquzzaman, M. 2003. "Error modeling schemes for fading channels in wireless: A survey." *IEEE Communications Surveys and Tutorials*, 5(2), 2-9.

[5] Di Marco, P.; Fischione, C.; Santucci, F.; Johansson, K. H. (2012). "Modeling IEEE 802.15.4 Networks over Fading Channels." *IEEE Transactions on Wireless Communications*, 13(10), 5366-5381.

[6] Ikram, W.; Jansson, N.; Harvei, T.; Aakvaag, N.; Halvorsen, I.; Peterson, S.; Carlsen, S.; Thornhill, N. F. 2014. "Wireless communication in process control loop: Requirements analysis, industry practices and experimental evaluation." *IEEE Emerging Technology and Factory Automation,* 1-8.

[7] Mazumder, S. K. 2011. *Wireless Networking Based Control*. Illinois: Springer.

[8] Colpo, J.; Mols, D. 2011. "No strings attached." In *Hydrocarbon Engineering*, 16(1), 47-52.

[9] Silva, I.; Guedes, L. A.; Portugal, P. 2013. "Emerging Technologies for Industrial Wireless Sensors. Networks." In *Embedded Computing Systems: Applications, Optimization, and Advanced Design*, 343-359, Hershey: IGI Global.

[10] Chang, K. D.; Chen, J. L. 2012. "A survey of trust management in WSNs, internet of things and future internet." *KSII Transactions on Internet and Information*, 6(1), 5–23.

[11] Gungor, V.; Hancke, G. 2009. "Industrial wireless sensor networks: challenges, design principles, and technical approaches." *IEEE Transactions on Industrial Electronics*, 56(10), 4258–4265.

[12] Han, S.; Zhu, X.; Chen, D.; Mok, A. K.; Nixon, M. 2011. "Reliable and real-time communication in industrial wireless mesh networks." *17th IEEE Real-Time and Embedded Technology and Applications Symposium*, 3–12.

[13] IEC 62591. 2010. *IEC 62591: Industrial communication networks – Wireless communication network and communications profiles – WirelessHART*. Geneva: International Electrotechnical Commission.

[14] IEC 62734. 2012. *IEC 62734: Industrial communication networks – Fieldbus specifications – Wireless systems for industrial automation: process control and related applications (based on ISA 100.11a)*. Geneva: International Electrotechnical Commission.

[15] Neves, P.; Rodrigues, J. 2010. "Internet protocol over wireless sensor networks, from myth to reality." *The Journal of Communication*, 5(3), 189-196.

[16] Soto, V. S.; Muller, I.; Winter, J. M.; Pereira, C. E.; Netto, J. C. 2014. "Control over WirelessHART through a host application: A WirelessHART Network Control Proposal." *Brazilian Symposium on Computer Systems Engineering*, 91-96.

[17] Tang, Z.; Zang, P.; Wang, H. 2010. "Analysis and design of real-time and reliable industrial wireless control communication network and

protocol." *Proceedings of 29th Chinese Control Conference*, 4157-4163.

[18] Zhang, W.; Braniy, M. S.; Phillips, S. M. 2001. "Stability of Networked Control Systems." In *IEEE Control Systems*, 21(1), 84-99.

[19] Song, J.; Mok, A. K.; Chen, D.; Nixon, M.; Blevins, T.; Wojsznis, W. 2006. *"Improving PID control with unreliable communications."* Paper presented at the ISA EXPO 2006, Houston, Texas, October 17-19.

[20] Werb, J.; Amidi, S. 2012. *"Control Over Wireless: Current Applications and Future Opportunities."* Paper presented at ISA Automation Week, Orlando, Florida, September 24-27.

[21] Ikram, W. 2012. *"Wireless Communication in Process Monitoring and Control."* PhD thesis, Imperial College London.

[22] De Biasi, M.; Snickars, C.; Landernas, K.; Isaksson, A. 2008. "Simulation of process control with WirelessHART networks subject to clock drift." *32nd Annual IEEE International Computer Software and Applications Conference*, 1355-1360.

[23] De Biasi, M.; Snickars, C.; Landernas, K.; Isaksson, A. 2008. "Simulation of process control with WirelessHART networks subject to packet losses." *IEEE International Conference on Automation Science and Engineering*, 548-553.

[24] Shah K.; Seceleanu T.; Gidlund, M. 2010. "Design and implementation of a WirelessHART simulator for process control." *International Symposium on Industrial Embedded System (SIES)*, 221-224.

[25] Ferrari, P.; Flammini, A.; Rizzi, M.; Sisinni, E. 2013. "Improving simulation of wireless networked control systems based on wirelessHART." *Computer Standards & Interfaces*, 35(6), 605-615.

[26] Wang, F.; Zeng, P.; Yu, H. 2008. *Slot time synchronization for TDMA-Based Ad Hoc Networks. International Symposium on Computer Science and Computational Technology*, 544-548.

[27] Hespanha, J. P.; Naghshtabrizi, P.; Yonggang, Xu. 2007. "A Survey on Recent Results in Networked Control Systems." In *Proceedings of the IEEE*, 95(1), 138-162.

[28] Bertelli, G.; Santos, A.; César, J.; Silva, I. 2016. "Performance Evaluation of ISA100.11a Wireless Feedback Control." *4th IFAC Symposium on Telematics Applications*, 49(30), 290-295.

[29] Ciccozzi, F.; Cicchetti, A.; Delsing, J.; Seceleanu, T.; Akerberg, J.; Carlsson, L. E. 2010. Integrating wireless systems into process industry and business management. *15th IEEE International Conference on Emerging Technologies and Factory Automation*, 1-8.

[30] Festo Didactic. *MPS-PA Compact Workstation*. User Manual.

[31] Dust-Networks. 2013. *DC9007A SmartMesh WirelessHART Starter kit*. Dust-Networks, Inc.

[32] Yang, T. C. 2006. "Networked Control Systems: A Brief Survey." *IEEE Proceedings on Control Theory Applications*, 153 (4), 403-412.

[33] Xin, K.; Xianghui, C.; Jiming, C.; Peng, C.; Lihua, X. 2015. Optimal controller location in wireless networked control systems. *International Journal of Robust and Nonlinear Control*, 25(2), 301-319.

[34] Demirel, B. 2015. *"Architectures and Performance Analysis of Wireless Networked Control Systems."* PhD thesis, KTH Royal Institute of Technology.

[35] Petersen, S.; Carlsen, S. 2009. "Performance Evaluation of WirelessHART for Factory Automation." *IEEE Conference on Emerging Technologies & Factory Automation*, 1-9.

[36] Park, P. G.; Araujo, J.; Johansson, K. H. 2011. "Wireless Networked Control System Co-Design." *International Conference on Networking, Sensing and Control*, 486-491.

[37] Nobre, M.; Silva, I.; Guedes, L. A. 2015. "Routing and Scheduling Algorithms for WirelessHART Networks: A Survey." *Sensors*, 15(5), 9703-9740.

[38] Saifullah, A.; Wu, C.; Tiwari, P.; Xu, Y.; Lu, C.; Chen, Y. 2012. "Near Optimal Rate Selection for Wireless Control Systems." *IEEE 18th Real Time and Embedded Technology and Applications Symposium*, 231-240.

[39] Winter, J. M.; Lima, C.; Muller, I.; Pereira, C. E.; Netto, J. C. 2011. "WirelessHART Routing Analysis Software." *Brazilian Symposium on Computing System Engineering*, 96-98.

[40] Kunzel, G.; Winter, J. M.; Muller, I.; Pereira, C. E.; Netto, J. C. 2012. "Passive Monitoring Software Tool for Evaluation of Deployed WirelessHART Networks." *Brazilian Symposium on Computing System Engineering*, 7-12.

[41] Han, S.; Zhu, X.; Mok, A. K.; Chen, D.; Nixon, M. 2011. "Reliable and Real-Time Communication in Industrial Wireless Mesh Networks." *IEEE Real-time and Embedded Technology and Applications Symposium*, 3–12.

[42] Pesonen, J. 2010. *"Stochastic estimation and control over WirelessHART networks: Theory and implementation."* MSc diss., KTH Royal Institute of Technology.

[43] Nobre, M.; Silva, I.; Guedes, L. A.; Portugal, P. 2010. *Towards a WirelessHART module for the NS-3 simulator. 5th Conference on Emerging Technologies & Factory Automation,* 1-4.

[44] Santos, A.; Lopes, D.; César, J.; Luciano, L.; Neto, A.; Guedes, L. A.; Silva, I. 2015. "Assessment of WirelessHART networks in closed-loop control systems." *IEEE International Conference on Industrial Technology*, 2172-2177.

[45] Rezah, F. P.; Shin, S. Y. 2013. "Performance Evaluation of ISA100.11a industrial wireless network." *IET International Conference on Information and Communication Technologies*, 587-592.

[46] Silva, I.; Guedes, L. A.; Portugal, P.; Vasques, F. 2012. "Reliability and availability evaluation of wireless sensor networks for industrial applications." *Sensors*, 12 (1), 806-838.

In: Networked Control Systems
Editor: Eduardo Paciência Godoy
ISBN: 978-1-53613-105-5
© 2018 Nova Science Publishers, Inc.

Chapter 4

OPTIMAL PLACEMENT OF ROUTERS IN INDUSTRIAL WIRELESS NETWORKS BASED ON MULTI-OBJECTIVES

***Anderson Costa Silva dos Santos, Daniel Lopes Martins, Heitor Medeiros Florencio, Jorge Dantas Melo and Adrião Duarte Dória Neto*[*]**
Department of Computer and Automation Engineering,
Federal University of Rio Grande do Norte, Natal, RN, Brazil

ABSTRACT

Nowadays wireless communication has aroused interest and has been a tendency in industry environments. In part this interest is due to the ease of deployment and maintenance of this network, which dispenses sophisticated designs and execution of works of deploying cables, which in industrial environment often prohibitively expensive, and enabling the addition of new applications when compared to their wired counterparts. Despite its high degree of applicability, a wireless network faces some technical (reliability, energy consumption, environment interference) and

[*] Corresponding Author address. Email: andersoncs@dca.ufrn.br.

cultural challenges present as the main negative points for the deployment of large-scale industrial networks in the industry. In this chapter will address the problem of equipment placement, routers as well as sensors and actuators, with focus on control in the industrial wireless networks, because the wireless equipment placement directly influences the number of routes and the instruments signal reach, enabling more powerful signal against obstacles and alternative routes in case of primary routes loss. With particular emphasis on the WirelessHART and ISA100.11a standard, which presents themselves as the most promising solution for industrial wireless networks, this chapter presents a multi-objective mathematical model to aid to the design and deployment of industrial wireless sensor and actuator network. The model consists of a search method based on the multi-objective genetic algorithm technique, taking into consideration criteria necessary for control such as packet delivery time and route stability, to determine the optimal routing instruments placement in order to the network can become more reliable and robust according to the design specifications of the desired industrial wireless environment.

Keywords: Industrial wireless network, WirelessHART, ISA100.11a, Multi-objective genetic algorithm

Introduction

Due to the non-deterministic wireless communication channel, companies have always been skeptical in the use of wireless technology. This scenario is aggravated by the fact that the equipment are installed in areas subject to the influence of external agents (noise, interference, adverse weather, natural obstacles), which can generate high transmission error rates when compared to cabled technologies. Signal attenuation (influenced by the distance between transmitter and receptor) and the problem of multiple paths (due to reflection, diffraction and spreading of the transmitted signal) might also cause other communication errors (Silva, Guedes and Portugal, 2013). Wireless communication errors are generally transient. Communication in the channel is noisy during a short period and then returns to normal (Willig et al., 2002). Unlike this technology, communication errors in wired networks are normally permanent and occur primarily due to failures in cables, connectors or other components (Marshall and Rinaldi, 2005).

Albeit those challenges, the evolution of wireless technology has been developing new mechanisms (modulation, coexistence with other technologies, cryptography, deterministic schedule, frequency hops and redundancy) to ensure the reliability of the network. The constant research allied to the creation of international standards, such as WirelessHART and ISA100.11.a, made the use of wireless technology viable in traditional industrial applications, such as in control system. These protocols enable the demand for new applications in industrial environments arise daily. The main investments are related with petrochemical plants, refineries, exploration, production and transportation of oil and gas (Hatler, Gurganious and Chi, 2008). Due to high probability of large financial losses in the occurrence of failures, reliability and availability assume important roles in the design of these applications (Talevski, Carlsen and Petersen, 2009). Assuming the use of wireless devices, other application requirements must also be analyzed, for example: operation in noisy environment or with obstructions, preventive maintenance, flexibility of configuration and installation, and integration with legacy IT solutions.

In addition, advances in the fields of wireless sensor networks (WSN) have created many new opportunities for research, where a control system routes its required signals via wireless communication links. Specific applications are emerging among urban utilities and for industrial plants that has large area of work, e.g., in public lighting systems, smart grids and oil and gas industry. Nowadays, the efforts are going toward modern communication systems with complex factors, such as size and heterogeneity and energy-saving solutions.

There is necessity of the industry to use these new communication technologies in critical scenarios and to control those systems. In addition, it is required to take into consideration that the performance of these systems depends on the performance of the equipment and this is directly related to communication among the devices. This work presents a multi-objective mathematical model to assist in the design and deployment of industrial wireless sensor and actuator network (WSAN). The model helps in the network forming by the optimal placement of routers aiming to improve the communication links among devices, which is based on the wireless sensor networks reliability.

The rest of this chapter is organized as follows: section 2 presents an overview of industrial wireless protocols; section 3 describes multi-objective optimization; the multi-objective mathematical model is shown in section 4; section 5 contains the testbeds. Finally, conclusions are presented in section 6.

Abbreviations and Acronyms

6LoWPAN:	IPV6 Over Low Power Wireless Personal Area Networks.
HART:	Highway Addressable Remote Transducer.
HCF:	HART Communication Foundation.
IEC:	International Eletrotechnical Commission.
IPV6:	Internet Protocol Version 6.
ISA:	International Society of Automation.
IO:	Input/Output.
IT:	Information Technology.
NSGA-II:	Non-dominated Sorting Genetic Algorithm II.
PAS:	Publicly Available Specification.
WSAN:	Wireless Sensor and Actuator Network.
WSN:	Wireless Sensor Network.

Industrial Wireless Network

The emergence of technologies for industrial wireless networks was a natural evolution of the legacy industrial communication technologies. The proposal to eliminate field wiring and use a new paradigm for data transmission in industrial environments is not recent. (Lessard and Gerla, 1988) developed one of the first works in this area in an attempt communicating industrial devices with infrared. According to (Colpo and

Mols, 2011), the use of wireless equipment can reduce installation costs by 50-90% compared to scenarios where wired devices are used.

Despite eliminating costs, industrial wireless networks still face many challenges. Some open issues are related to addressing, routing, managing devices with limited physical capabilities (energy, processing, memory, etc.), security and privacy, dealing with heterogeneous technologies, safety, and standardization. Other relevant issues to be analyzed are dependability requirements (reliability and availability), as faults may lead to system failures, which can result in financial losses, environmental damage, or putting people in danger.

Due to those challenges and the demand for using wireless technology, wireless communication standards were developed in order to fulfill this need, with WirelessHART and ISA.100.11.a being the two main technologies used in industrial environments. The ISA.100.11.a standard focuses on process control and monitoring applications, with latency requirements of around 100 ms.

WirelessHART

WirelessHART is a standard of wireless communication developed by the HART Communication Foundation (HCF) aimed at transmitting HART messages without using classic means of transmission (4-20 mA or RS484). A WirelessHART device implements the same structure of commands used by a classic HART RS484 device. The same applications used in the HART standard are compatible with the WirelessHART standard. Its final version (IEC 62591) was approved in 2010 (IEC, 2010).

The WirelessHART specification defines 8 types of devices, as described in Figure 1: network manager, security manager, gateway, access point, field device, adapter, router and portable device. Among all the supported devices, the network manager is considered the most important. Normally found installed in gateway, the network manager is responsible for controlling all network events (routing, scaling, alarms, etc).

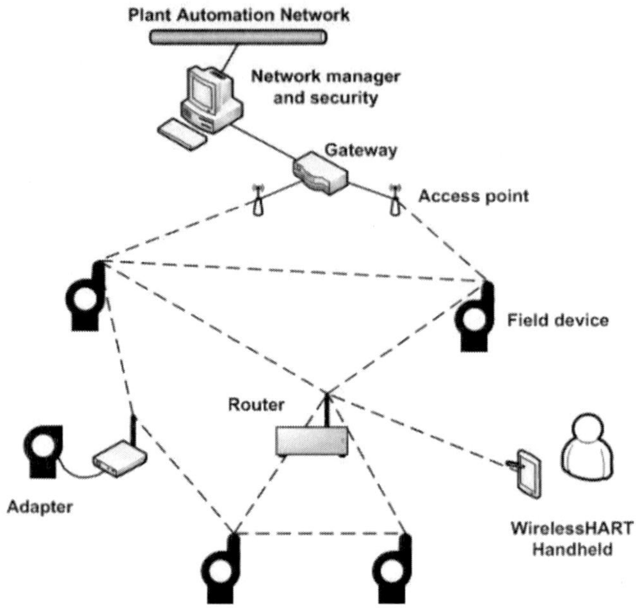

Figure 1. WirelessHART Devices.

Routing Aspects

A WirelessHART network supports mesh topologies in order to create redundant paths between the devices and the gateway. In this regard, WirelessHART specifications define four types of routing: source routing, graph routing, superframe routing and proxy routing. Both of them use information collected by devices and transmitted to the network manager.

The network manager creates routing and the devices use the configurations assigned to them (Silva et al., 2012). Because only the system manager has the entire view of the network and manager routes, listening to the other to aiming to find any alive device and what is the best neighbor with the best signal could increase the time to send or resend packets and waste a large quantity of energy of each instrument.

This is the reason because field devices periodically send information from the network to the manager. Whenever possible the network manager configures each transmission with at least two routing options (primary and secondary) in order to increase the network reliability.

Moreover, to minimize the influence of interference (noise) in the network and allow coexistence with other standards (IEEE 802.11, Bluetooth, ZigBee), a frequency hopping mechanism was added to the routing procedure (data link layer). Each transmission in the network uses a different frequency (channel), thereby decreasing the likelihood of choosing a noisy channel.

Network Formation

The network manager initiates the WirelessHART network formation process, creating a secure and reliable communication channel with the gateway. Initial provisioning from the access point will occur through this channel, and consists of transmitting the following information:

- Superframe management, which guarantees a minimum bandwidth to execute basic network monitoring and maintenance functions;
- Network graph for upstream traffic, that is, towards the network manager;
- Superframe and join links, which enable the entry of new devices into the network;
- Dedicated and shared links (both transmission and reception) to manage the devices, health reports traffic and alarms communication (lost links, for example).

The network formation process is divided into three stages: advertisement, join and parameter negotiation. Field devices that have already undergone the join process carry out advertisement, and its function is to inform the presence of the network and allow the incorporation of new devices within its communication range. The packet of advertisements must contain network ID and ASN information (Absolute Slot Number), in addition to join and superframe links. This packet is expected for passive join devices, whose response will be in the form of a request for the device in question to join the network, that is, a join request.

As soon as a new device is added to the network, both the gateway and the device will make bandwidth requests to the network manager. The gateway needs bandwidth to support request/response traffic, while the device will use the necessary bandwidth to publish information on the process variable, for example. The network manager will estimate the necessary bandwidth using these service requests, thereby optimizing the network manager. If there is insufficient bandwidth, the manager will allocate superframes and links according to the requests. Otherwise, the manager can allocate less bandwidth than the device requested, or even refuse the request (Rech, 2012).

ISA100.11a

In 2009, the International Society of Automation (ISA) approved a wireless mesh-networking standard known as ISA100.11a. In 2011, the International Electrotechnical Commission (IEC) created a task force to approve ISA100.11a as a publicly available specification. Currently, the ISA100.11a standard is named in the draft version as IEC/PAS 62734 (IEC-62734, 2012). Differently from the WirelessHART standard, whose development was based on a specific technology (HART), the ISA100.11a standard was designed based on the user application requirements (ISA100, 2009).

The main goals of the ISA100.11a standard are related to guaranteeing the deterministic communication latency while increasing the communication reliability. It focuses on process control and monitoring applications, with latency requirements of around 100 ms. ISA 100.11a can coexist with other wireless technologies such as cell phones, IEEE 802.11, IEEE 802.15, and IEEE 802.16, and can provide tunneling for legacy protocols (HART, Foundation Fieldbus, Profibus, and Modbus).

A typical ISA 100.11a network, which is described in Figure 2, may be composed of seven types of devices: gateway, system manager, security manager, router, backbone router, input/output (IO) devices, and portable devices. In addition, more than a role could be assigned to one device. The IO device is responsible for monitoring the environment and it has no routing function. This limitation was configured to minimize the energy consumption. An IO device can also be portable, and in this case, its operation is limited to testing, management, and configuration of other devices.

Two devices conduct the routing mechanism: the router and the backbone router. The former works as a proxy, forwarding the network data towards the gateway, whereas the latter is responsible for high level routing only between the gateway and the backbone routers.

All traffic between the backbone routers is based on 6LoWPAN (IPV6 over low power wireless personal area networks). Additionally, the backbone routers can segregate the network into sub-nets. Similarly, to the WirelessHART network, an ISA100.11a network is centralized to three devices: the gateway, security manager, and system manager (Neves and Rodrigues, 2010).

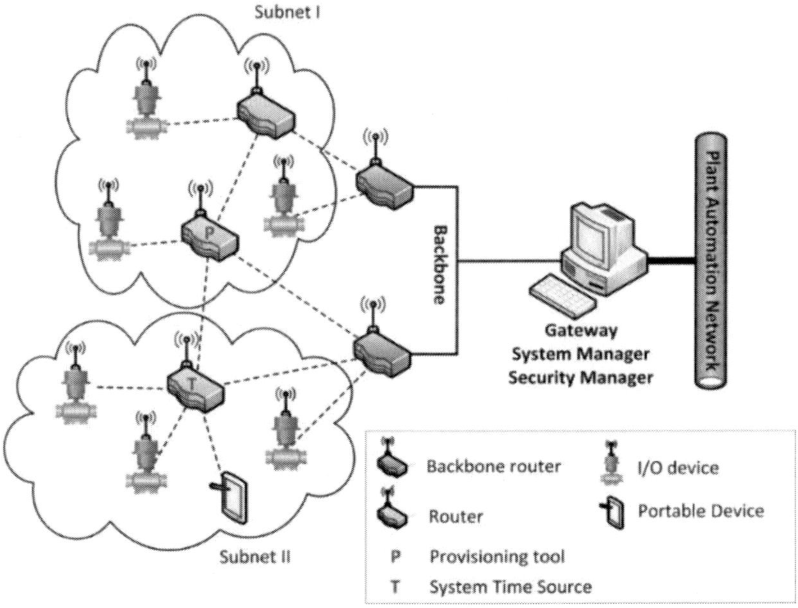

Figure 2. A typical ISA100.11.a Network.

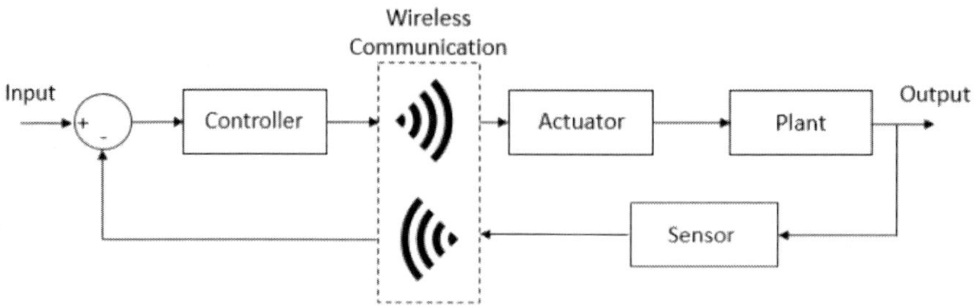

Figure 3. Wireless Feedback Control.

The gateway device provides a connection between the wireless sensor network and the plant automation network. It is also responsible for guaranteeing the tunneling with the legacy technologies. Due to its importance, there is support for redundant gateways. The security manager and the system manager perform the most important tasks. The security manager controls the system security management function whereas the system manager governs all the network, devices, and communications. ISA.100.11.a standard supports multiple packets reception sequentially (duocast).

Although the ISA.100.11.a technology has proven to be reliable, the control systems in process industry can be quite complex due to need low latency and high reliable. Therefore, given how familiar traditional control techniques are, companies tend to be very reluctant to invest the time needed to apply wireless measurement and control elements to control.

Wireless Control

A wireless feedback control system consists of the basic closed-loop control system in which the communication is done via a wireless channel. The measured output signal that is feedbacked to the controller as well as the controller signal applied to the actuator are sent through a wireless network. Figure 3 illustrates the system.

MULTI-OBJECTIVE OPTIMIZATION

Multi-objective optimization considers optimization problems involving more than one objective function to be optimized simultaneously. Multi-objective optimization problems arise in many fields, such as engineering, economics, and logistics, when optimal decisions need to be taken in the presence of trade-offs between two or more conflicting objectives (Miettinen, 2008). For example in wireless networks, rising distance among wireless devices enforcing to rise power signal (that is a goal) to enable communication, but also rising energy consumption (that is a problem) enforcing industrial plants stop for maintenance resulting in injury.

Typically, there does not exist a single solution that simultaneously optimizes each objective. Instead, there is a (possibly infinite) set of Pareto optimal solutions. A solution is called non-dominated or Pareto optimal if none of the objective functions can be improved in value without degrading one or more of the other objective values. Without additional subjective preference information, all Pareto optimal solutions are considered equally good (Fonsecay and Flemingz, 1993).

In mathematical terms, a multi-objective optimization problem can be formulated using (1).

$$\min\big(f_1(x), f_2(x), \dots, f_k(x)\big) \qquad (1)$$
$$s.t. x \in X$$

Where the integer $k \geq 2$ is the number of objectives and the set X is the feasible set of decision vectors. The feasible set is typically defined by some constraint functions. In addition, the vector-valued objective function is often defined as $f: X \rightarrow R^k$, $f(x) = (f_1(x), f_2(x), \dots, f_k(x))^T$. An element $x^* \in X$ is a feasible solution; a feasible solution $x^1 \in X$ is said to (Pareto) dominate another solution $x^2 \in X$, if

$$f_i(x^1) \leq f_i(x^2), \text{ for all indices } i \in \{1, 2, \dots, k\} \text{ and} \qquad (2)$$
$$f_j(x^1) < f_j(x^2), \text{ for at least one index } j \in \{1, 2, \dots, k\}.$$

In this work, the search for the Pareto solution set has been made using multi-objective genetic algorithms explained below.

Genetic Algorithm

The genetic algorithm is a very effective technique to search optimal solution for optimization problems. It starts with an initial population ($P_t = 0$) of solutions encoded as shown in (3). An initial population is most often generated randomly but a heuristic can used. A chromosome represents each population individual, which is made of sequence of genes and every gene controls the inheritance of specific attributes of the solution characteristics. A fitness function in line 4 and 9 measures the quality of chromosome in terms of various design variables of the solution. A more fitted chromosome suggests a best individual. The while loop in lines 4 up to 9 represents the evolution process, where relatively fit individuals reproduce new designs and inferior design die. This process continues until a design with desirable fitness is found. Line 6 selects the best designs within the current generation based on fitness values. These selected designs known as parents are used to reproduce the next generation of designs known as offspring. The evolution process involves two genetic operations namely mutation in line 7 and cross over line 8. A mutation operator are arbitrarily alters one or more genes of randomly selected chromosome. The intuition behind the mutation operator is to introduce a missing feature in the population (Lanza-Gutierrez and Gomez-Pulido, 2015).

A cross over operator combines the features of two selected chromosomes (parents) to form two similar chromosomes (offspring) by swapping genes of parent chromosome. The intuition behind the cross over operator is to exchange information behind different potential solution.

In this work, mutation was used to evolve the population of the genetic algorithm.

$$
\begin{aligned}
&1. t = 0 \\
&2. Initialize\ P_t \\
&3. Evaluete\ P_t \\
&4. while\ (termination\ conditions\ are\ unsatisfied) \\
&5. t = t + 1 \\
&6. Select\ \mathrm{P}_t\ from\ \mathrm{P}_{t-1} \\
&7. Mutate\ some\ of\ \mathrm{P}_t \\
&8. Crossover\ some\ of\ \mathrm{P}_t \\
&9. Evaluate\ \mathrm{P}_t
\end{aligned}
\tag{3}
$$

NSGA-II

NSGA-II is a popular non-domination based genetic algorithm for multi-objective optimization. It was chosen to find optimal solution for proposed multi-objective mathematical model because it is a very effective algorithm and has low computational complexity. NSGA-II was developed, which has a better sorting algorithm, incorporates elitism and no sharing parameter needs to be chosen a priori (Lanza-Gutierrez and Gomez-Pulido, 2015).

The population is initialized as usual. Once the population in initialized the population is sorted based on non-domination into each front. The first front being completely non-dominant set in the current population, the second front being dominated by the individuals in the first front only, and the front goes so on. For each individual in each front are assigned rank (fitness) values or based on front in which they belong to. Individuals in first front are given a fitness value of 1 and individuals in second are assigned fitness value as 2 and so forth.

In addition to fitness value, a new parameter called *crowding distance* is calculated for each individual. The crowding distance is a measure of how close an individual is to its neighbors. Large average crowding distance will result in better diversity in the population.

Parents are selected from the population by using binary tournament selection based on the rank and crowding distance. An individual is selected in the rank if it is smaller than the other or if crowding distance is greater than the other.

A new population formed from the previous population and current offspring is sorted again based on non-domination and only the best *N* individuals are selected, where *N* is the population size. The selection is based on rank and the crowding distance based on the rank from the first rank until the last. When the entire rank could not be inserted on the population, then in this case, the individuals are chosen by crowding distance.

NSGA-II can be summarized to the following steps:

$$
\begin{aligned}
&1. Create\ parents\ |P_0| = N\ (random)\\
&2. Create\ Q_0 = \emptyset\ (offspring)\\
&3. From\ t = 0\ until\ maxGen\\
&4. R_t = P_t \cup Q_t\\
&5. fronts\ F = fast\_non\_dominated(R_t)\\
&6. P_{t+1} = \emptyset\ and\ i = 0\\
&7. Until\ |P_{t+1}| + |F_i| \leq N\\
&8. calculate\ crowding\_distance(F_i)\\
&9. P_{t+1} = P_{t+1} + F_i \qquad (4)\\
&10. \mathrm{i} = i + 1\\
&11. P_{t+1} = Sort_{\preccurlyeq n}(F_i)\ (descending\ order)\\
&12.\ P_{t+1} = P_{t+1} \cup F_i[1\colon (N - |P_{t+1}|)]\\
&13. Until\ |Q_{t+1}| \leq N\\
&14. Selection\ (fathers\ f_1\ and\ f_2\ from\ P_{t+1})\\
&15. Crossover\ (f_1\ and\ f_2\ \Rightarrow\ s_1\ and\ s_2)\\
&16. Mutation\ (over\ s_1\ and\ s_2)\\
&17. Q_{t+1} = Q_{t+1} \cup s_1 \cup s_2
\end{aligned}
$$

MULTI-OBJECTIVE MODEL

With the goal of proposing the optimal placement of routers in WSANs that has control in the industry, this section will present the multi-objective mathematical model to aid to the design and deployment of industrial wireless sensor and actuator network to control. This model is based on packet propagation in wireless channels via disjoint routes, increasing network communication reliability.

WirelessHART and ISA100.11a standards have a lot of similarities, but also differences. Therefore, standardize the industrial wireless networks for testbeds is needed.

Industrial Wireless Network Standard

For both protocols:

- An industrial wireless network contains a set of field devices (sensors, actuators and routers), a gateway, an access point and a network manager. Many manufacturers produce their gateway containing access point and network manager, therefore in this chapter, gateway will use as general device that contains those 3 equipment.
- The industrial environment is a two dimensional area or plane.
- All the information captured by the sensors (information for neighbors, links, signal power for each neighbor, battery, process variable) are forwarded to the gateway.
- Each instrument has a primary and a secondary route determined by WirelessHART and ISA100.11a standards. If its primary route fails, device sends packets over its secondary route. If the primary route will not be available again, gateway allows secondary route becomes primary route and choices a new secondary route.
- Network can contain up to 100 devices (manufacturer gateway limitation).
- Sensors also are considered routers in both protocols (WirelessHART and ISA100.11a).
- If sensor is used for only routing, that is its process variable is not considered in control process, it is considered only router.

- The network topology can be mesh or star, according to device signal reach. To determine the reach, taking into consideration the transmit power (it cannot exceed 10 dBm) and reception threshold (considering -85 dBm based on IEEE 802.15.4 standard).
- The controller will be on a computer connected to the gateway through the automation network due to gateway limitation.

Signal Propagation

To know if each device is receiving communication packets is needed to calculate its reception signal level. For each device is calculated reception signal level through path loss using (5) (Chan et al., 2007).

$$PL = PL_0 + 10 * n * \log_{10}\left(\frac{d}{d_0}\right) + X \quad (5)$$

Where:

- PL is path loss;
- PL_0 is path loss to base distance;
- n is path loss exponent (it equals to 2);
- d_0 is base distance (in this work, 1 m);
- X is normal or Gaussian random variable of mean 0 referring to attenuation.

Therefore, reception signal level is calculated using (6).

$$P_{Rx_{dBm}} = P_{Tx_{dBm}} - PL \quad (6)$$

PL_0 could be obtained from average of measurements from the site at the base distance, that is 1 m, and for these tests, 49 dBm was obtained as path loss.

Disjoint Routes

A network is called k ($k \geq 2$) disjoint vertices if for each pair of devices in network there are k disjoint routes. In addition, a disjoint path is a path that does not share devices with another path. Thus, no device loses communication in a network k disjoint vertices if they loss communication with less than k neighbors (Han et al., 2010).

A path or route is a mean where there is communication between gateway and sensor or actuator. That can be direct or indirect (through routers).

Figure 4 shows disjoint routes example. F1 has 3 disjoint routes (F1-R1-G, F1-R2-G e F1-R3-G) e others non-disjoint routes (F1-R1-R2-G has R2 as well as F1-R2-G).

In this work, it was adopted each device must have 3 disjoint routes (one primary, one secondary and extra route for use if primary route fails and it will not be available again).

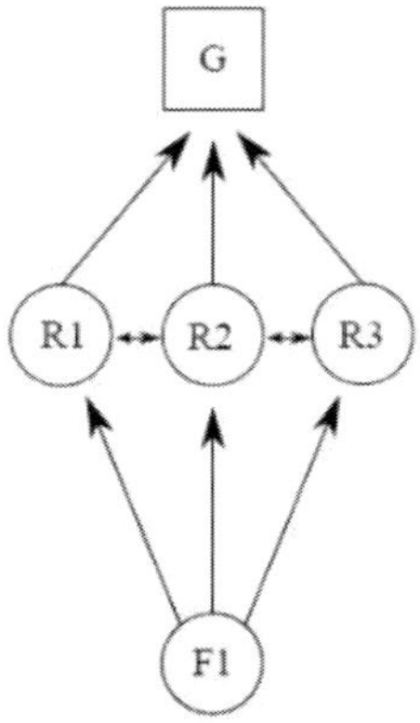

Figure 4. Disjoint routes.

Disjoint Routes among Instruments

Disjoint routes among instruments are defined as the number of routes that do not share devices on paths to the gateway. Therefore, given 2 devices

F1 e F2, they have 2 disjoint routes if there are 2 routes between F1 and gateway and between F2 and gateway that do not have common instruments. Figure 5 illustrates this approach.

That displays that F1 and F2 have at least 2 disjoint routes among them (F1-R1-G, F1-R2-G and F2-R4-G, F2-R5-G) e 1 shared route between both (F1-R3-G and F2-R3-G). The greater the amount of routers increases the possibility of there are disjoint routes among neighbor instruments and decreases possibility of communication fails in network, increasing network reliability.

Average Reception Power

Due to some factors as electronic components, antenna, interference and even signal obstruction, signal level can changes on network paths. Average reception power can be calculated using (7).

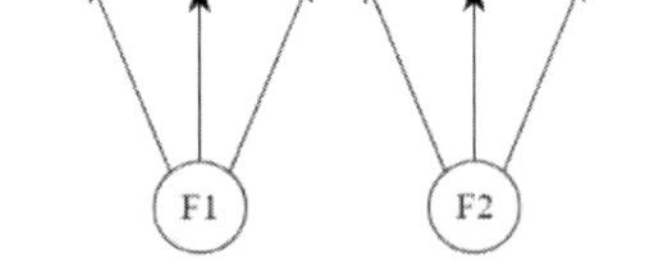

Figure 5. Disjoint routes among instruments.

$$mean\ P_{Rx_{1n}} = \frac{\sum_{i=2}^{n} P_{Rx_{i \to i-1}} + P_{Rx_{i-1 \to i}}}{2 * (|n| - 1)} \qquad (7)$$

Where n is nth instrument and $|n|$ is amount of instruments on route starting from the instrument 1 up to nth instrument.

Multi-Objective Mathematical Model

Albeit a wireless network can have many instruments, it is not sufficient to result in a lot of diversity of routes, and disjoint routes enables to use duocast feature in wireless network, that allows multiple packets reception sequentially, increasing packet delivery ratio, increasing reliability of the network link communication. For example, a network can contain many routes with same router; if this router fails, many routes are affected generating many faults in network. The length of the routes influences packet delay. If routes have many hops, consequently, there are delays on control process.

The average power is also an important item due to influence signal quality and reach. High average power increases reception signal level and decreases possibility of communication failures on control process.

The amount of routers on network is a factor to be analyzed due to high acquisition and maintenance cost making control process impossible.

Analyzing the behavior of wireless networks according to placement of its devices, it is perceived these concerns are influenced by placement of devices in network. Thus, this work proposes a multi-objective mathematical model to aid to the design and deployment of industrial wireless sensor and actuator network to control. The model consists of a search method based on the multi-objective genetic algorithm technique, taking into consideration criteria necessary for control such as packet delivery time and route stability, to determine the optimal routing instruments placement in order to the network can become more reliable and robust according to the design specifications of the desired industrial wireless environment. In this case, the packet delivery time was analyzed based on how long is the route (number of routers).

Given the importance of criteria above, the model uses these criteria as objective functions in order to find an optimal placement of routers in WSANs to enable control in industrial wireless network.

Defining M_{LR} as the average length of routes throughout the network, it can be estimated using (8).

$$M_{LR} = \frac{\sum_I \frac{\sum_{R_i} SR_{ir}}{|R_i|}}{|I|} \tag{8}$$

Where I is set of all sensors on network, R_i is set of routes of specific sensor, SR_{ir} denotes the number of instruments in route r of instrument i and |I| is size of set.

Considering mDR as the minimum number of disjoint route combinations of all the network instruments, it can be express for (9).

$$mDR = \min(DRQ_i, \forall i \in I) \tag{9}$$

While DRQ_i is the number of combinations N in 3 of routes of instrument i that do not have common instruments (excluding sensor/source and gateway/destination) and it is shown in (10).

$$DRQ_i = \sum_{C_i} Comb_C \tag{10}$$

C_i is set of combinations of routes of instrument i and $Comb_c$ is a function that checks if there are not common routes, defined by (11).

$$Comb_C = \begin{cases} 1, if \ \cap_3 C_{R_r} = \emptyset \\ 0, otherwise \end{cases} \tag{11}$$

Where C_{R_r} is set of instruments of route r of combination$c$.

For different instruments, there is also a similar logic to ensure each instrument has n disjoint routes among all devices.

The number of combinations of network disjoint routes is given for (12).

$$DRNQ = \sum_{CN} Comb_n \tag{12}$$

CN is the set of combinations between routes of each network instrument and routes of the other instruments and $Comb_n$ is the function defined by (13).

$$Comb_n = \begin{cases} 1, if \ \bigcap_{|n|} n_{R_i} = \emptyset \\ 0, otherwise \end{cases} \quad (13)$$

Where |n| is the size of the set of combination n of routes of network instruments, n_{R_i} is the set of instruments of route i of combinationn.

Thus, to define the multi-objective optimization problem, the objective functions is shown in (14).

$$f_1 = M_{LR}, f_2 = mDR, f_3 = DRNQ, f_4 = |R_i|, f_5 = meanP_{Rx_{1n}} \quad (14)$$

Where RI is the set of routers and $\mathrm{meanP}_{\mathrm{Rx}_{1n}}$ is the average power defined by (5) and multi-objectives are defined in (15).

$$\min(f_1), \max(f_2), \max(f_3), \min(f_4), \max(f_5) \quad (15)$$

Subject to (16).

$$|R_i| \geq 3, \forall i \in I, mDR > 1, DRNQ > 1 \quad (16)$$

The gene encoding used for the genetic chain was binary. Each router is represented by a bit on genetic chain and to change routes (and the network architecture) just need to active or not (change from 1 to 0) a router.

Experimental Results

In this section, experimental results are presented. To test the concept, an ISA100.11a network was built. The first step was to configure wireless devices and consequently the network. NSGA-II parameters were configured in sequence, creating the testbed scenario.

The industrial environment is a two dimensional area (width: 1,590.75 m; height: 1,228.50 m). Routers are allocated up to 100 (manufacturer gateway limitation) and in random placements.

NSGA-II is configured by the following parameters:

- Population size is 50 individuals;
- 2 stopping condition was considered:
  - Maximum number of generation is 4000.
  - Percentage of population that does not exchange information behind different potential solution, it is 50%.
- Tournament table size is 3;
- Probability of cross over is 95%;
- Probability of mutation is 1%.

After the configuration, the multi-objective optimization problem defined in (13) and subject to (14) was used to find optimal placements of routers in WSANs.

According to the need of WirelessHART and ISA100.11a standard, each sensor/actuator must have at least 2 routes (primary and secondary). Ordinarily, the primary route is used to send packets, but when there is failed to send packets by primary route, secondary route is chosen to send packets in order to avoid packet loss. If a primary route device has failed to send packets during a long time, the gateway routing algorithm changes the primary route device to be the secondary route device and that chooses a new secondary route device in available routes set, in addition to the that updates all network device route tables. This procedure aims to increase the stability in packet delivery and avoid packet loss. Thus to increase the stability in packet delivery is need to increase number of routers (increasing routes per device), but this generates slowness in gateway data processing (updating network device route tables), harming communication on network links. To allow the exchange of routes in case of failures on network, avoiding harming communication on network links, each sensor/actuator has a combination of 3 disjoint routes to gateway. In addition at least 1 disjoint route among neighbors is required, avoiding many devices route loss when only one router fails, since the router is not shared by others paths, increasing network reliability.

Taking into consideration the multi-objective model and industrial environment and the NSGA-II configuration, the developed program aims to find a network topology. Each sensor/actuator has a combination of 3 disjoint routes to gateway and each sensor/actuator has at least 1 disjoint route among neighbors that do not share devices on paths to the gateway with others sensors/actuators to improve control communication link stability, decreasing packet loss. The initial environment is shown in Figure 6.

The program allows each instrument has one primary route and one secondary route. For example, in Figure 6, TT-IS-03 and TT-IS-01 has TT-IS-05 on their primary routes (non-disjoint routes) to gateway and they have disjoint secondary routes. Face that, NSGA-II is used to search optimal placements of routers on defined industrial environment. While initial devices (sensors and actuators) are fixed on environment for plant control, routers could be allocated in many places aiming to improve network reliability and consequently control reliability on wireless network.

The search consists to evaluate possible placements of routers evaluating its routes combinations aiming to ensure multi-objectives of model. Observing this problem involves more than one objective function to be optimized simultaneously and some objectives are opposites.

On this scenario, routers are dispersed aiming to maintain distance at least 1 meter from another devices. Table 1 shows a small part of the experimental results obtained for the proposed algorithm, where TI is the total number of network instruments, TR is the total number of network routers, RDI is the number of combinations of 3 disjoint routes for each instrument and RDN is the number of combinations of 1 disjoint route among instruments.

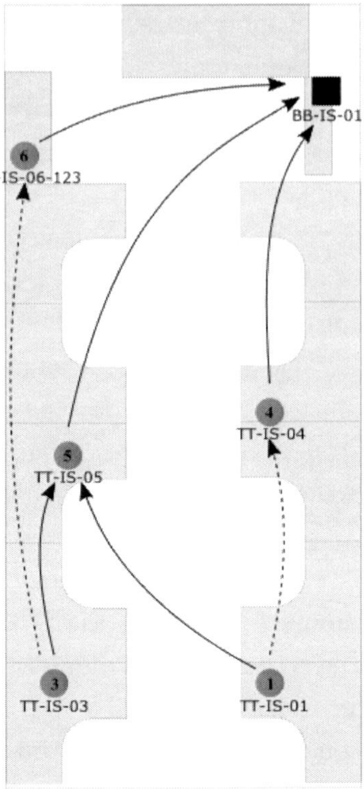

Figure 6. Initial environment.

Table 1. Variables

TI	TR	RDI	RDN
10	4	1	15
20	14	9	861
31	25	220	942
42	36	120	952
52	46	455	969
66	60	1771	967
71	65	2300	970

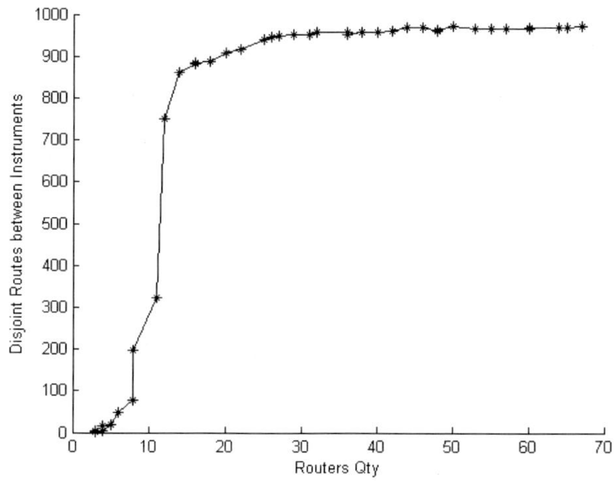

Figure 7. Pareto optimal solutions (Routers Quantity x Disjoint Routes between Instruments).

The obtained results generate set of Pareto optimal solutions and none of the objective functions can be improved in value without degrading one or more of the other objective values. The set of Pareto optimal solutions is a multidimensional surface. Based on the set of Pareto optimal solutions, we can define the best solution according to any available parameter on the mathematical model.

The algorithm searches for a combination of placement of routers that has at least 1 disjoint route among neighbors, therefore the graph does not start at 0, and quantity of routers is minimum possible to achieve that goal. As the number of routers grows, new combinations of routes are possible for the network, increasing a quantity of disjoint routes among instruments, but the growth is limited because the algorithm normally does not analyze all combinations of disjoint routes, it analyzes according to route length, sorting the routes in ascending order by length. The first routes analyzed to combinations of disjoint routes are those that have 1 router between sensor/actuator and gateway. If there are not at least 3, disjoint routes with 2 routers between sensor/actuator and gateway are analyzed to complete at least 3 disjoint routes. If there are not at least 3, disjoint routes with 3 routers are analyzed and so forth. Therefore that process searches for 3 disjoint routes with minimum length, albeit the placement of routers is random and it does not change during the analysis process, the routes do not change, consequently, quantity of combinations of disjoint routes also do not change.

Figure 7 illustrates a small part of the set of Pareto optimal solutions by number of routers and number of combinations of disjoint routes among instruments. The number of disjoint routes rises (following number of routers) until 30 routers when there are few modifications because the algorithm does not change routes and it finds the set of optimal solution.

Figure 8 shows Pareto optimal solutions by the number of routers and the number of combinations of disjoint routes by instruments. The number of disjoint routes by instruments rises following number of routers. As graph in Figure 7, the graph in Figure 8 takes into consideration at least 1 disjoint route among neighbors and at least 3 disjoint routes by instruments. If the number of routers grows, the number of combinations of disjoint routes also grows. This graph is obtained using (9), the minimum number of disjoint route combinations of all the network instruments, that is, insert routers in network may alter disjoint route by instrument or not because it is used minimum number of combinations. Thus, to change the minimum number of disjoint route combinations of all the network instruments it is need to insert a number of routers that increase number of disjoint route for all instruments simultaneously. It explains the slower growth of the second graph and why even with more routers, at some points in the graph, the number of combinations of disjoint routes by instrument is less that other points with less routers. At these points chosen routers may have increased the number of routes of an instrument, but not all.

After first testbed, a second scenario with more instruments was analyzed. Second scenario represents a scenario found in an industry and it is shown in Figure 9.

All instruments (except 5) have at least 3 neighbors and they are distributed that clusters of instruments are formed to meet industry demand. In Figure 10, the behavior of graph in Figure 7 is repeated. As the number of routers grows, increasing a quantity of disjoint routes among instruments until a limit. It is observed in this scenario that some routers influence directly the number of disjoint routes among instruments until 30 routers, after 30 this sensitivity is hardly observed.

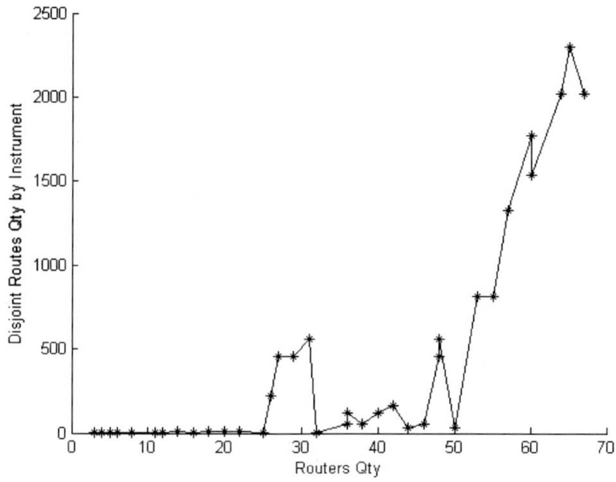

Figure 8. Pareto optimal solutions (Routers Quantity x Disjoint Routes by Instrument).

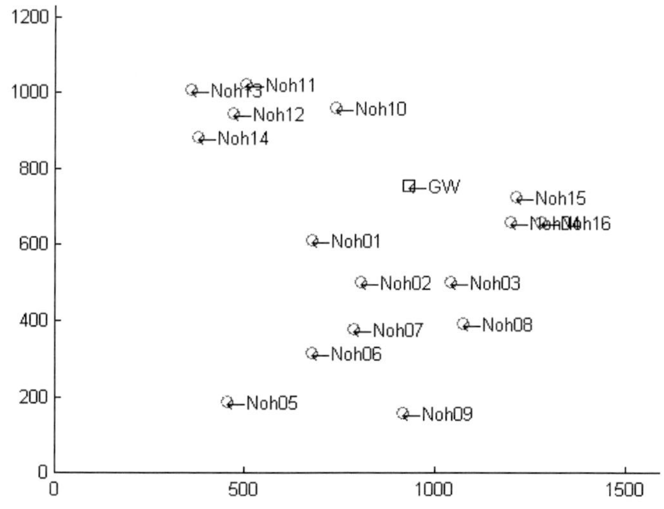

Figure 9. Second scenario.

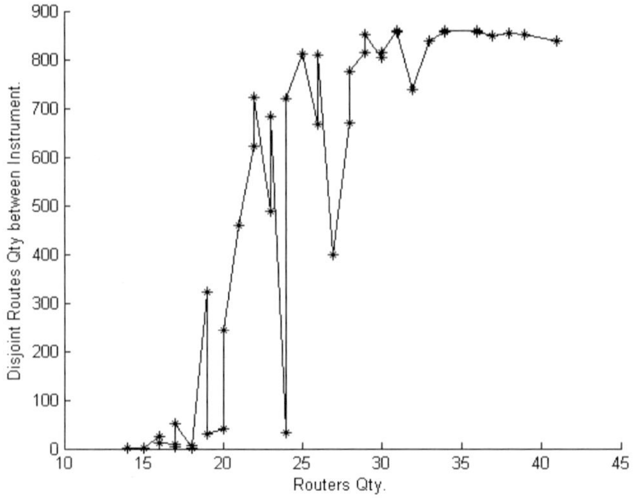

Figure 10. Pareto optimal solutions for second scenario.

As the graph of the number of disjoint routes by instrument in initial environment, the number of combinations just grows if the number of combinations grows for all instruments in second scenario. Thus, the addition of routers does not imply growth at the same time, because what matters most is where the routers are located. It is observed when 23 routers are inserted, increasing the minimum number of routes, differently when 24 routers are inserted.

In Figure 11, it is shown Pareto optimal solutions by the number of routers and the number of combinations of disjoint routes by instruments. For this scenario, there are communication bottlenecks among clusters, therefore the change of few routers alter the number of combinations of disjoint routes by instrument. If a router is inserted within a cluster, there is no modification on the number of disjoint routes, but if a router is inserted outside a cluster, it rapidly increases the number of disjoint routes by instrument. It also explains why there is no increase in the number of routes in relation to the number of routers. As the instruments are randomly allocated currently, the optimal placement of routers (it would be inserted near the bottlenecks) is not always found.

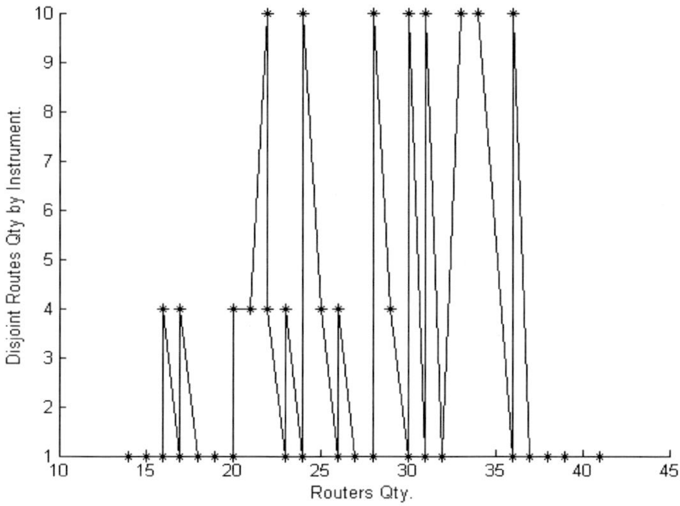

Figure 11. Pareto optimal solutions for second scenario.

CONCLUSION

In this chapter, it was described the main industrial wireless sensor and actuator network technologies. The features described in the chapter can be used to aid the designer to develop networks that are more resilient. From the standards mentioned, it is possible to note that all they are based on the IEEE 802.15.4 standard.

The multi-objective mathematical model proposed has main objective to decrease number of wireless network instruments and route lengths, increasing reliability of network link communication. Besides disjoint routes enables to use duocast feature in wireless network, in other words, it allows multiple packets reception sequentially, decreasing packet loss.

Experimental results for automatic router placement problem was presented. The main objectives are length of routes, using disjoint network routes and average signal power to improve control communication links. The experiments with multi-objective mathematical model to aid to the design and deployment of industrial wireless sensor and actuator network to control demonstrate very promising results. In future, it will consider more sophisticated network models.

ACKNOWLEDGMENTS

The authors thank CNPq (National Research Council) for financial support in the form of a research scholarship and the UFRN (Federal University of Rio Grande do Norte) for the support infrastructure to conduct this work.

REFERENCES

Chan, T. M., K. F. Man, K. S. Tang, and S. Kwong. 2007. "A jumping-genes paradigm for optimizing factory wlan network" In *IEEE Transactions on Industrial Informatics*, 33-43.

Colpo, J., and D. Mols. 2011. "No strings attached" In *Hydrocarbon Engineering*, 47-52.

Fonsecay, Carlos, and Peter Flemingz. 1993. "Genetic algorithms for multiobjective optimization: Formulation, discussion and generalization" In *Proceedings of the Fifth International Conference*, 1-8.

Han, X, X. Cao, E. L. Lloyd, and C. C. Shen. 2010. "Fault-tolerant relay node placement in heterogeneous wireless sensor networks" In *IEEE Transactions on Mobile Computing*, 643-656.

Hatler, Mareca, Darryl Gurganious, and Charlie Chi. 2008. "Wireless sensor networks for oil & gas" In *Technical report*, On World.

IEC. 2010. "IEC 62591: Industrial communication networks" In *Wireless communication network and communications profiles - WirelessHART*.

ISA100. 2009. "ISA100.11a - Wireless systems for industrial automation: Process control and related applications" in *ISA Standards*.

Lanza-Gutierrez, Jose M., and Juan Gomez-Pulido. 2015. "Assuming multiobjective metaheuristics to solve a three-objective optimization problem for relay node deployment in wireless sensor networks" In *Soft Computing*, 675-687.

Lessard, A., and M. Gerla. 1988. "Wireless communications in the automated factory environment" In *Network, IEEE*, 64-69.

Marshall, P., and J. S. Rinaldi. 2005. "ISA: The Instrumentation, Systems, and Automation Society" In *Industrial Ethernet*, 2 edition.

Miettinen, Kaisa. 2008. "Introduction to Multiobjective Optimization: Noninteractive Approaches" In *Multiobjective Optimization,* edited by Jürgen Branke, Kalyanmoy Deb, Kaisa Miettinen, and Roman Słowiński, 1-26. Spinger.

Neves, P. and J. Rodrigues. 2010. "Internet protocol over wireless sensor networks, from myth to reality" In *Journal of Communications*, 189-196.

Rech, Jônatas Romani. 2012. *"Development of a WirelessHART network manager",* 20-21.

Silva, Ivanovitch, Luiz Affonso Guedes, and Paulo Portugal. 2013. "Emerging Technologies for Industrial Wireless Sensor Networks" In *Embedded Computing Systems: Applications, Optimization, and Advanced Design*, edited by Mohamed Khalgui, Olfa Mosbahi, and Antonio Valentini, 343-359. IGI Global.

Silva, Ivanovitch, Luiz Affonso Guedes, Paulo Portugal, and Francisco Vasques. 2012. "Reliability and Availability Evaluation of Wireless Sensor Networks for Industrial Applications", *Sensors*, 806-838.

Talevski, A., S. Carlsen, and S. Petersen. 2009. "Research challenges in applying intelligent wireless sensors in the oil, gas and resources industries" In *Industrial Informatics, INDIN*. 7th IEEE International Conference on, 464-469.

Willig, A., M. Kubisch, C. Hoene, and A. Wolisz. 2002. "Measurements of a wireless link in an industrial environment using an IEEE 802.11-compliant physical layer" In *IEEE Transactions on Industrial Electronics*, 43:1265-1282. Accessed July 28, 2017.

In: Networked Control Systems
Editor: Eduardo Paciência Godoy
ISBN: 978-1-53613-105-5
© 2018 Nova Science Publishers, Inc.

Chapter 5

THE STUDY OF COMMUNICATION BETWEEN DISTRIBUTED GENERATION DEVICES IN A SMART GRID ENVIRONMENT

Jefferson Aparecido Dias**, Paulo José Amaral Serni and Eduardo Paciência Godoy
São Paulo State University (UNESP), Sorocaba, SP, Brazil

ABSTRACT

The increasing use of electricity due to the population growth in recent years has resulted in a necessary expansion of the energy matrix through alternative sources, including microprosumers in low or medium voltage. Therefore, studies have been conducted to use the automation to control systems, but there is difficulty in establishing what will be controlled and how to get the data effectively. In the Microgrids scenario, one of the challenges is the network communication process between different devices, such as generators controllers, smart meters, etc., in order to facilitate the information exchange and improve network performance. Therefore, this chapter presents a study of a Microgrid (MG) with a

* Corresponding author address: Email: jeffdias2000@yahoo.com.br.

Distributed Generator (DG) controlled remotely via a wireless communication. The methodology is presented by smart grid simulations, where the electrical system was implemented using the software PLECS – blockset, the wireless network was developed using TrueTime – toolbox, both emulated in Matlab®/Simulink®. In this application of MG, it was used a WLAN and ZigBee network to control the generation of the active power of and is evaluated under different network parameters, such as the loss probability and distance between devices. The simulation results show that the effectiveness of networked solution for developing MG. The results show that the simulation system provided a solution for the communication networks analysis, the DG control and modeling of electrical network in the smart MG, showing the effectiveness of the network solution for the development of MG.

Keywords: distributed generation, smart grids, wireless network, integrated simulation, smart system

INTRODUCTION

A Smart Grid (SG) is an electricity network that can intelligently and automatically integrate the activities of all users connected to it such as generators, consumers and those that do both in order to efficiently deliver sustainable, economic and secure electricity supplies (Vijayapriya and Kothari, 2011). The increasing use of electricity in recent years has demanded the adoption of policies that would encourage the use of low and medium power distributed generation - DG technologies, including renewable energy or alternative sources and energy storage elements (El-hawary, 2014). Considering the several challenges in changing the electricity network, microgrids have emerged as an economically and feasible option. A microgrid (MG) is generally known as a local and small-scale grid consisting of distributed generating stations along with consumer loads (Mahmoud, Azher Hussain and Abido, 2014). In MGs, generating stations are usually designed in a way the power can directly flow from the generators to the loads without having to pass through the transmission network. MGs also support a flexible and efficient electric grid, by enabling the integration of renewable energy sources such as solar, wind, and facilitating the deployment of DGs. Moreover, the use of local energy sources attending local loads supports reducing energy losses in transmission and distribution, providing efficiency to the electric (Guerrero et al., 2010).

Regardless of having some benefits of MG, there are some challenges related to this such as implementation needs, security and control requirements among others. One of the main issues in discussion for the SG development is the communication among the devices to be installed in the environment, such as smart meters, energy storage systems, management and control systems, users and the DGs (IEEE, 2011; Gungor et al., 2011 and Souza et al., 2015). Different communication technologies can be used for different purposes and requirements of each specific smart grid application. There are several papers addressing communication technologies, requirements, architectures and standards for SGs (Gungor et al., 2011 and Gungor et al., 2013). An extensive revision about communication technologies and standards is present in (Gungor et al., 2011), compiling the benefits of wired and wireless solutions for different applications. In (Gungor et al., 2013), proposes an architecture for SGs consisting of three layers: application layer, the power layer, and the communication layer. A description about these layers along with a discussion about its communication requirements such as latency, data rate, security and reliability is presented. In (Sabbah, El-Mougy and Ibnkahla, 2014), discusses about some challenging aspects related to the architectural model for SGs proposed by the National Institute of Standards and Technology (NIST, 2014). This paper also focus on the analysis of routing protocols for smart grids, providing detailed review of existing protocols and a comparison between them.

In view of the SG scenario, modeling and simulation tasks are also challenging but promising. The integration among different areas creates a tradeoff between reliable scenarios and results with the difficult to develop them. In (Mahmoud, Azher Hussain and Abido, 2014), presents an overview of recent developments in modeling and control methods of MGs, including a detailed description about microgrids and its modes of operation. This revision papers attests that the MG modeling and control depends on the components used for the microgrid (DG unit, energy-storage unit, controller unit and conventional load) and discusses recent techniques available on literature. A distributed networked control is used in order to implement a secondary control in a cascaded controlled MG is presented in (Shafiee, Guerrero and Vasquez, 2014). Even though this paper considers the networked MG, there is no concern with the type and dynamics of communication link used, which can be seen as an optimistic assumption on the literature of networked control systems. A simplified fixed communication latency link is used in the experiments.

In this context, co-simulation tools that allows the integrated modeling and simulation of communication networks, electrical systems and control strategies in the same environment are of key importance for the development of comprehensive smart microgrid testbeds, instead of (or before) their experimental implementation. Therefore, this paper presents a preliminary simulation study of a simplified smart MG with DG modeled in the PLECS blockset (Plecsim, 2002); controlled through a WLAN and a ZigBee wireless network modeled using the TrueTime toolbox. Both tools are integrated in Matlab/Simulink in which the control system was designed.

Proposed Smart Microgrid

These are modern, small-scale versions of the centralized electricity system that aims to achieve specific local targets such as CO_2 reduction, cost reduction and diversification of energy matrices. The smart MG, as well as the conventional energy grid, generates, distributes and regulates the flow of electricity to consumers locally, and is an ideal way to integrate renewable resources at community level and enable customer participation in the electricity company, forming the building blocks of the perfect power. In this section, we present the methodological and material aspects for the integrated simulation of the proposed electric circuit containing distributed generation, power generation control and the wireless communication network, as well as the tools used for the generated set to be simulated in the Matlab-Simulink environment.

The electric circuit with the DG is presented in simplified form in Figure 1.

The circuit of Figure 1 is composed of a single-phase voltage source and a line impedance (Z), which resembles a real network. The load is of the resistive type with nominal power. The respective values of voltage and current of the grid and current in the DG, are measured at the common coupling point (PCC), which allows to observe the behavior of the load. A dashed line between the main network and the load with the represents the PCC. A DG with capacity to generate power of up to 10 kW was inserted in parallel with the load. However, in order to analyze the behavior of the MG with the DG and the control system, the interaction value was set at 5 kW and measured by the voltage reference to the current generated in the PCC.

The objective of the DG is to generate fundamental active power (P_1) which, in this case, will be consumed by the load, regardless of the waveform of the source voltage and current.

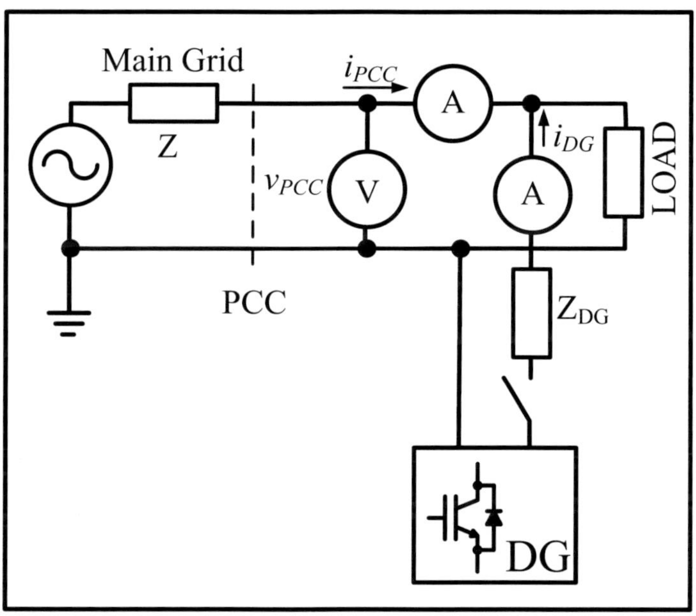

Figure 1. Diagram of the electric circuit with the DG.

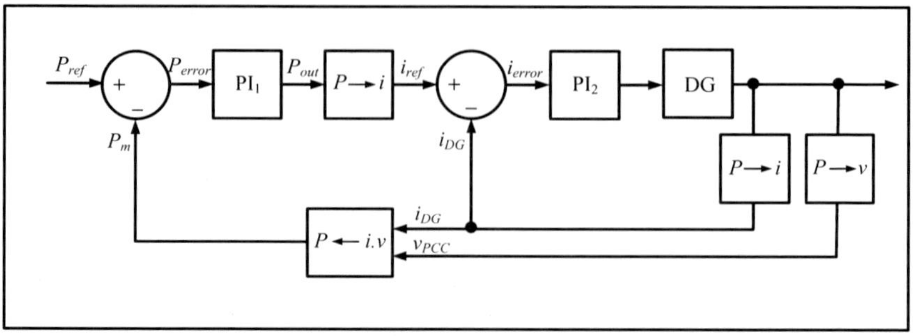

Figure 2. Control of the active power generation of DG.

Figure 2 shows the power generation control for the DG, where the error between the reference power (P_{ref}) and the measured power (P_m), generated by the DG, passes through a controller of the integral proportional type (PI_1) resulting in the value of the output power (P_{out}) of the controller.

The power P_m is calculated as (1).

$$P_m = \frac{1}{T}\int_0^T v_{PCC}^1\, i_{DG}^1 dt \qquad (1)$$

where: v^1_{PCC} and i^1_{DG} are respectively the plots of fundamental instantaneous voltage in the PCC and fundamental instantaneous current of the DG.

The reference instantaneous current i_{ref} for the DG is obtained according to (2).

$$i_{ref} = \frac{P_{out}}{(V_{PCC}^1)^2} v_{PCC}^1 \qquad (2)$$

where: V^1_{PCC} is the RMS value of v^1_{PCC}, which can be obtained according to (3).

$$V_{PCC}^1 = \sqrt{\frac{1}{T}\int_0^T (v_{PCC}^1)^2\, dt} \qquad (3)$$

Finally, the error between i_{ref} and i^1_{DG} passes through another PI_2 controller going to the DG, which in the simulations operates with an output sine voltage.

In this application, active power generation is controlled remotely over a wireless communication network. Figure 3 shows the proposed communication network, as well as the exchange of information between each device.

The Node 1 device can be a microcontroller or microprocessor specialized in digital signal processing (DSP), responsible for processing the measured data from the current and voltage sensors, to calculate the P_m, control the current and send signals to the investor's communication in DG. In this case, Node 1 is composed of the Sensor-Actuator set, where the sensor receives the values of v_{PCC} and i_{DG} and through this data calculates P_m, which will be sent via wireless to Node 2, controller. Node 2 is responsible for controlling the active power, receiving P_{ref} and P_m from Node 1 via wireless network and sending P_{out} to Node 1 via wireless network. Node 1 receives the power P_{out} through the actuator and sends to the DG the i_{DG} signal to be injected into the network. Additional functions can be directed to Node 2, such as supervision control, user interface and communication with utilities or smart meters.

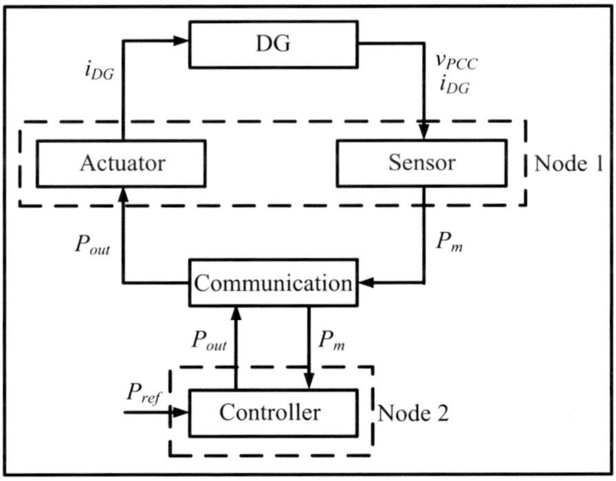

Figure 3. Proposed wireless network.

Thus, the exchanges of information between Node 1 and Node 2 are, respectively, the values of P_m and P_{out}, which can be transmitted with a sampling rate of at least the fundamental frequency of the network, i.e., 60

Hz enabling analyze the behavior of the system with each cycle. Therefore, in order to obtain the power value, at least one sampling cycle is necessary, since the frequency of the electric network is also 60 Hz.

Figure 5 shows the circuit of a smart MG in a simplified way with the identifications and separations as to the type of software used in the development of each step of the circuit.

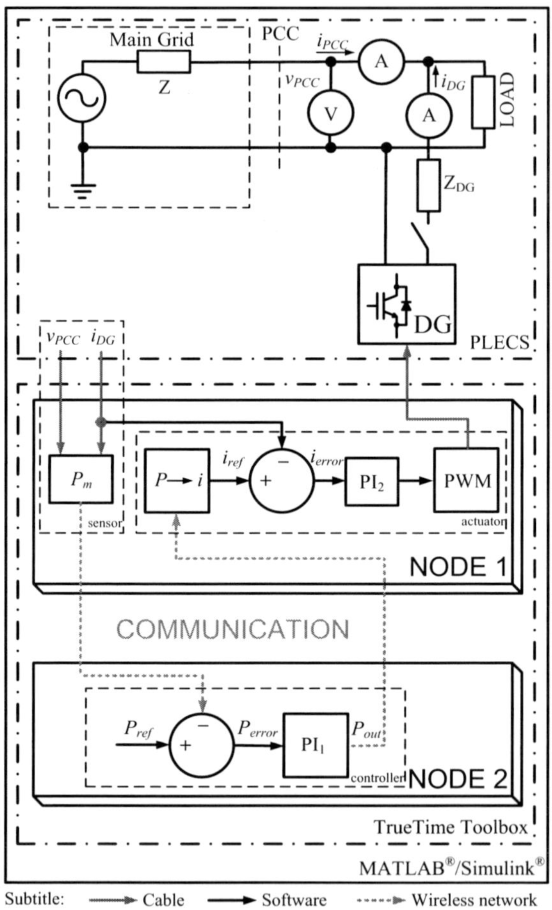

Figure 4. Circuit of an MRI.

Materials and Methods

The wireless network was simulated using the TrueTime-toolbox software in Matlab-Simulink environment and the blocks are described in the following subsections.

TrueTime Wireless Network

The wireless network block is based on events and simulates the MAC mechanism, the transmission of the message in the network and the processing time, according to the type of network and protocol chosen (Cervin, Ohlin and Henriksson, 2007).

Each message transmitted on the network contains information about the source and target device of the message, user-defined data such as measurements performed, control signals, and message size. Some specific information about real-time characteristics such as priority and deadline messages are optional and defined according to the protocol chosen (Godoy, 2011).

This block is used to represent a wireless network, allowing access to the transmission medium and data packet, being configured in this work for the following types of network protocols:

- IEEE 802.11b/g ($WLAN$);
- IEEE 802.15.4 ($ZigBee$).

Figure 5 shows the TrueTime Wireless Network block, where:

- The values 0 and 20, are constants that represent the distance between Node 1 and Node 2 that in this case was 20 meters;
- Schedule is used to check the behavior of the network during data transmission;

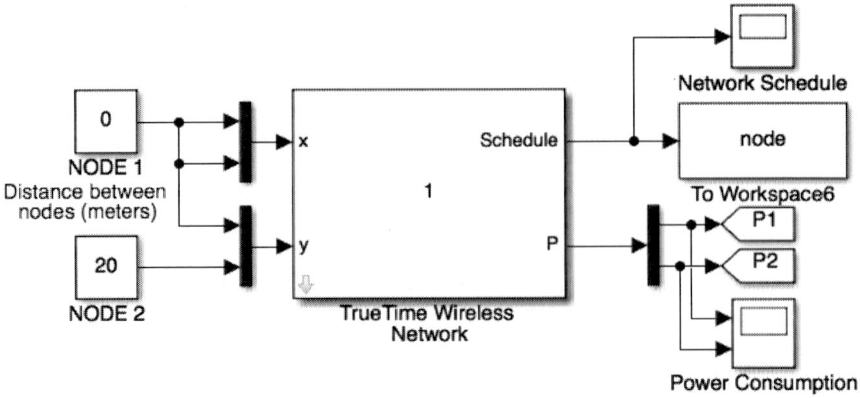

Figure 5. TrueTime Wireless Network Block.

P1 and P2 refers to the energy consumption in the communication between Node 1 and Node 2 respectively.

TrueTime Kernel

The Kernel Block consists of a Matlab function that simulates a computer or a processor with an event-based kernel, A/D and D/A conversion operations, I/O data interface, network communication, external channel for interrupts and monitoring variables (Schedule and Monitors) (Cervin, Ohlin and Henriksson, 2007).

It is used to represent the node in the network, that is, the component or device on which the network will exchange information.

For each component or device connected to the network, it is necessary to use such a block.

Figure 6 shows the TrueTime Kernel block and the TrueTime Battery block used in Node 1.

Figure 7 shows the TrueTime Kernel block and the TrueTime Battery block used in Node 2.

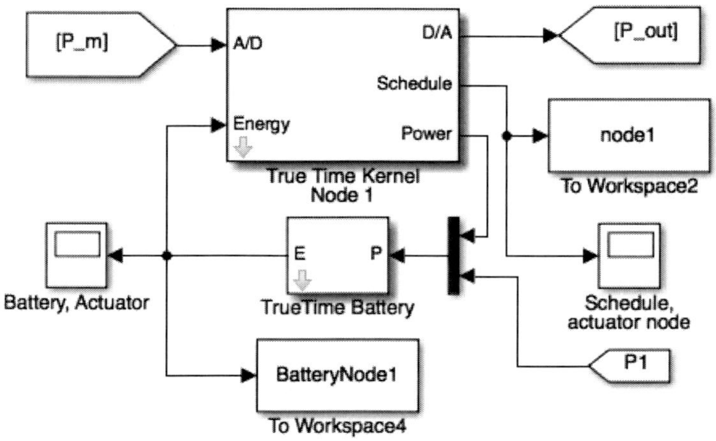

Figure 6. TrueTime Kernel block and TrueTime Battery block of Node 1

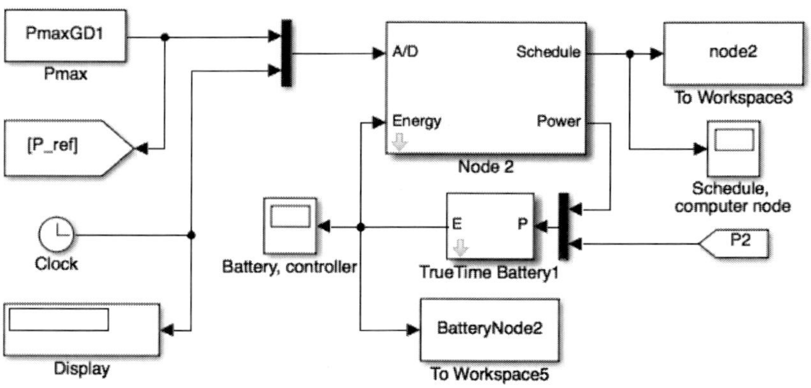

Figure 7. TrueTime Kernel block and TrueTime Battery block of Node 2.

where:

- A/D is the analog data input that is converted to digital data;
- Energy is the connection point that is connected to the E connection of the battery;
- D/A is the output where the digital data is converted to the analog data;

- Schedule is the location for checking the behavior data of the respective network node;
- Power is the connection point that is connected to the P connection of the battery.

TrueTime Battery

The True Time Battery block was developed with the purpose of allowing the simulation of control via the network in a mobile robot environment, in which batteries supply the power supply of the devices.

Figure 6 and Figure 7, respectively, present the TrueTime Battery block, where:

- P is the connection point that is connected to the Power connection of the TrueTime Kernel block;
- E is the connection point that is connected to the Energy connection of the TrueTime Kernel block.

RESULTS AND ANALYSIS

The voltage supplied by the main network is 127 V_{RMS} with a frequency of 60 Hz. The grid impedance values are represented by an equivalent Z, where R = 23.8 mΩ and L = 104 μH. The value of Z can be obtained from (4), (5) and (6) respectively, or according to (7).

$$Z = R + j \cdot X \tag{4}$$

$$X = \omega \cdot L \tag{5}$$

$$\omega = 2 \cdot \pi \tag{6}$$

$$Z^2 = R^2 + X^2 \tag{7}$$

With the impedance value, it was possible to calculate the value of the angle and the short-circuit current, according to (8) and (9).

$$tan\theta = \frac{X}{R} \tag{8}$$

$$I_{CC} = \frac{V}{Z} \tag{9}$$

The total simulation time was 10 s and assuming for that time interval that the maximum power that can be injected by the DG was 5 kW, P_{ref} was also set at 5 kW. In addition, the DG was connected at time t = 0.1 s to analyze the behavior of the instantaneous voltage and current in the PCC and instantaneous current in the DG, as well as the behavior of the powers through the wireless communication.

Figure 8 shows P_{PCC}, P_{ref} and P_m.

Figure 9 shows the waveforms of v_{PCC} and i_{PCC} in the PCC while Figure 10 shows i_{DG} in the first few seconds of simulation. Note that the currents are in phase with the voltage and this is because the load is resistive.

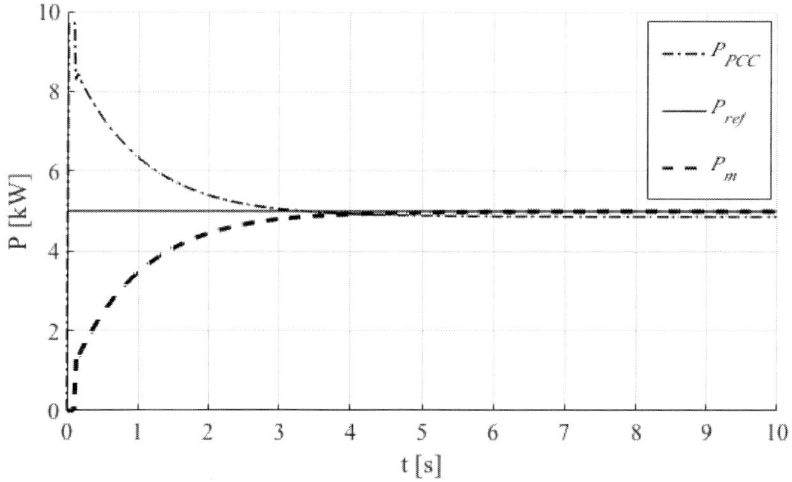

Figure 8. P_{PCC}, P_{ref} and P_m.

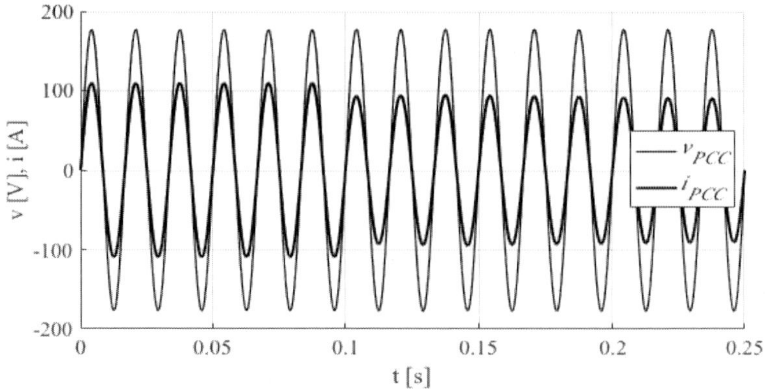

Figure 9. v_{PCC}, i_{PCC}.

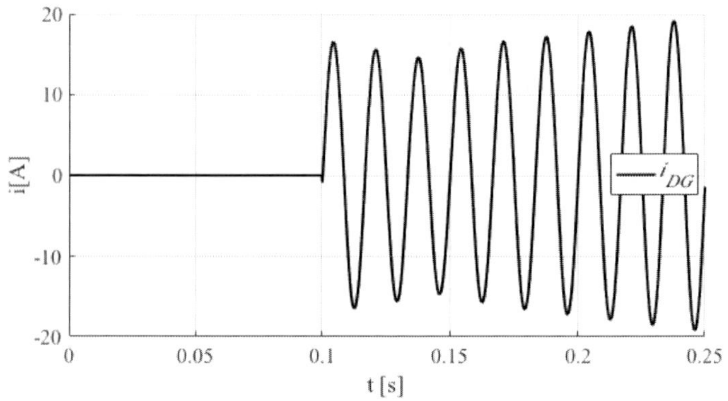

Figure 10. iDG.

The oscillations in the amplitude of the waveforms of v_{PCC} and i_{PCC} occur due to the transient period.

The analysis was performed for two types of wireless network protocols, as described below:

Protocol IEEE 802.11b/g (WLAN)

For this analysis, it was considered an ideal data transmission network, that is, with 0% probability of data loss throughout its transmission.

Table 1. Configuring the block truetime wireless network

Parameters			
Item	Description	Data	Unit
1	Network type	802.11b/g	-
2	Network Number	1	-
3	Number of nodes	2	-
4	Data rate	64000000	bits/s
5	Minimum frame size	272	bits
6	Transmit power	13	dbm
7	Receiver signal Threshold	-71	dbm
8	Pathloss function	default	-
9	Pathloss exponent	3,5	1/distance
10	ACK timeout	0,00004	s
11	Retry limit	5	-
12	Error coding threshold	0,03	-
13	Loss probability (0–1)	0,0	-
14	Initial seed	0	-

In the proposed smart MG circuit, the IEEE 802.11b/g (WLAN) network protocol was used in the TrueTime Wireless Network block and configured according to (Gigi, 2016) Table 1.

In addition to the settings presented in Table 1, the following options were selected:

Table 2. Truetime kernel block configuration of the node 1

Parameters		
Item	Description	Data
1	Name of init function (MEX or Matlab)	sensor_actuator_init
2	Init function argument (arbitrary struct)	[ ]
3	Number of analog inputs and outputs	[1 1]
4	Number of external triggers	0
5	(Network and) Node Number(s)	1
6	Local clock offset and drift	[0 0]

- *Show Schedule output port.*
- *Show Power consumption output port.*

Node 1 uses the TrueTime Kernel block and is configured according to Table 2.

Table 3 presents the configuration parameters block of the Node 2.

In addition to the settings presented in Tables II and III respectively, the following options were selected:

- *Show Schedule output port.*
- *Show Energy supply input port.*
- *Show Power consumption output port.*

The configuration parameter of the TrueTime Battery block of Node 1 and Node 2 respectively refers to the Initial energy field, which in spite of not having a unit of measure in the block, is in Joule, whose value entered was 2694.6 according (Cervin, Henriksson and Ohlin, 2010).

The value of the battery's power was based on a real power source composed of three rechargeable AA-type Nickel Metal Hydride - NiMH batteries, each with a voltage of 1.2 V, with a nominal load of 2300 mAh with high current capacity and considered discharged when its voltage is equal to 1 V (Thezolin et al., 2010).

Table 3. Truetime kernel block configuration of the node 2

Parameters		
Item	Description	Data
1	Name of init function (MEX or Matlab)	controller_init
2	Init function argument (arbitrary struct)	[ ]
3	Number of analog inputs and outputs	[1 0]
4	Number of external triggers	0
5	(Network and) Node Number(s)	2
6	Local clock offset and drift	[0 0]

The connected batteries in series, total a voltage of 3.6 V since the module of WLAN network works between 3.14 V to 3.46 V (Digi, 2016).

The battery power data can be obtained according to (10).

$$P_n = \frac{mAh \cdot V_{CC}}{1000} \tag{10}$$

where:

- Pn is the rated power in hours whose unit is in Wh.

To transform Wh to Joules, to enter the correct value in the Initial energy field of the TrueTime Battery block, we have (11).

$$E = W \cdot s \tag{11}$$

The difference between Ei and Ef can be obtained according to (12).

$$\Delta E = E_i - E_f \tag{12}$$

Figure 11 shows details of the use of the wireless network with their respective nodes during the simulation.

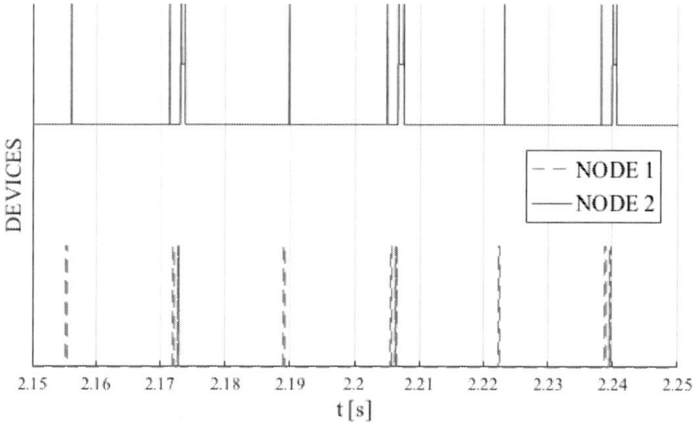

Figure 11. Details on using the wireless network with the smart MG Nodes (Node 1 - Sensor/Actuator and Node 2 - Controller).

Table 4 shows the processing of the network, of Node 1 and Node 2 respectively with respect to sending or receiving data and communication test, in percentage divided into three times, that is:

Table 4. Result of integrated smart mg simulation with 20 m distance and no loss of data on transmission (WLAN)

	Wireless Network Use (%)	Node 1 Use (%)	Node 2 Use (%)
t_{ON}	1,65	12,75	5,04
t_{OFF}	91,86	79,00	94,17
t_{WAIT}	6,50	8,25	0,80

- t_{ON} - Time the network is transmitting.
- t_{OFF} - Time the network is not transmitting.
- t_{WAIT} - Time the network is waiting for a priority node to finish transmitting so that it returns the transmission that had been started.

It can be observed that all data transmission in the network remained idle for most of the simulation time. This information is important due to the fact that the consumption of the devices is strictly t_{ON}-related together with the t_{WAIT} of the tasks.

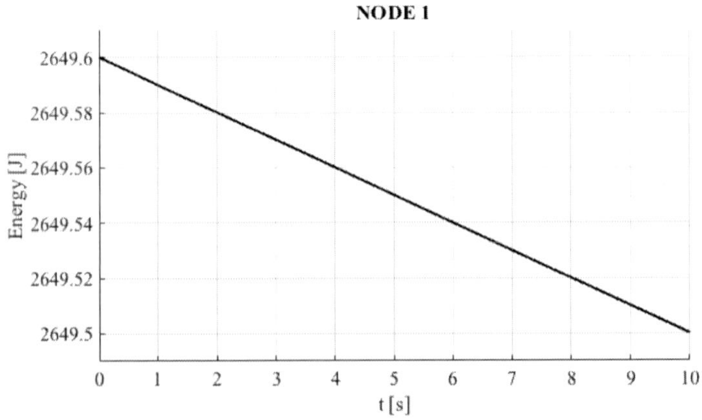

Figure 12. Battery consumption behavior in Node 1.

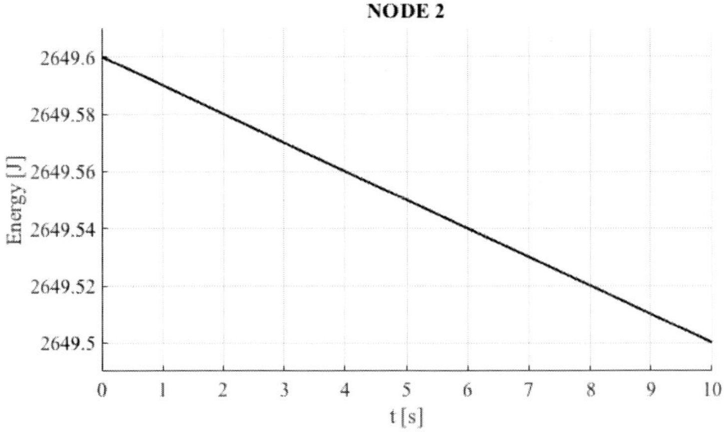

Figure 13. Battery consumption behavior in Node 2.

Figure 12 and Figure 13 shows the consumption behavior of the batteries during data transmission between the nodes.

Protocol IEEE 802.15.4 (ZigBee)

For this analysis, it was considered an ideal data transmission network, that is, with 0% probability of data loss throughout its transmission.

In the proposed smart MG circuit, the IEEE 802.15.4 (ZigBee) network protocol was used in the TrueTime Wireless Network block and configured according to Table 5. Due to the type of protocol, only item 1 has been changed which refers to the type of network, item 4 which refers to the size of the message to be transmitted, item 5 which refers to the transmission of energy and item 6 which refers to the sensitivity regarding the reception of the information, according (Digi, 2016)

The configurations of the Node 1 and Node 2 blocks did not change, remaining with the values according to Tables II and III.

The configuration parameter of the TrueTime Battery block of Node 1 and Node 2 respectively refers to the Initial energy field, which despite not having a unit of measure in the block, is in Joule, whose value entered was 4968, according (Cervin, Henriksson and Ohlin 2010).

The batteries connected in series, have a voltage of 3.6 V since the ZigBee network module operates between 2.7 V to 3.6 V (Digi, 2016 - 2).

Table 5. Configuring the block truetime wireless network

Parameters			
Item	Description	Data	Unit
1	Network type	802.15.4	-
2	Network Number	1	-
3	Number of nodes	2	-
4	Data rate	250000	bits/s
5	Minimum frame size	272	bits
6	Transmit power	18	dbm
7	Receiver signal Threshold	-101	dbm
8	Pathloss function	default	-
9	Pathloss exponent	3,5	1/distance
10	ACK timeout	0,00004	s
11	Retry limit	5	-
12	Error coding threshold	0,03	-
13	Loss probability (0–1)	0,0	-
14	Initial seed	0	-

The battery power data can be obtained according to (10), (11) and (12).

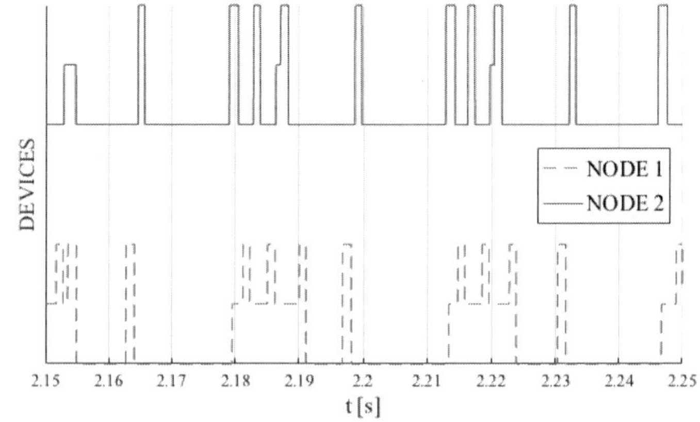

Figure 14. Details on using the wireless network with the nodes in the MRI (Node 1 - Sensor/Actuator and Node 2 - Controller).

Table 6. Results of the integrated smart mg simulation with 20 m distance and no loss of data in transmission (ZigBee)

	Wireless network Use (%)	Node 1 Use (%)	Node 2 Use (%)
t_{ON}	27,12	11,35	3,77
t_{OFF}	48,03	84,33	96,23
t_{WAIT}	24,85	4,32	0

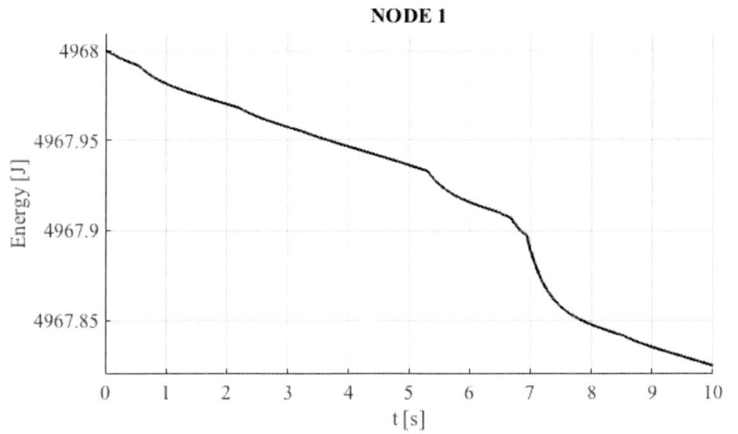

Figure 15. Battery consumption behavior in Node 1.

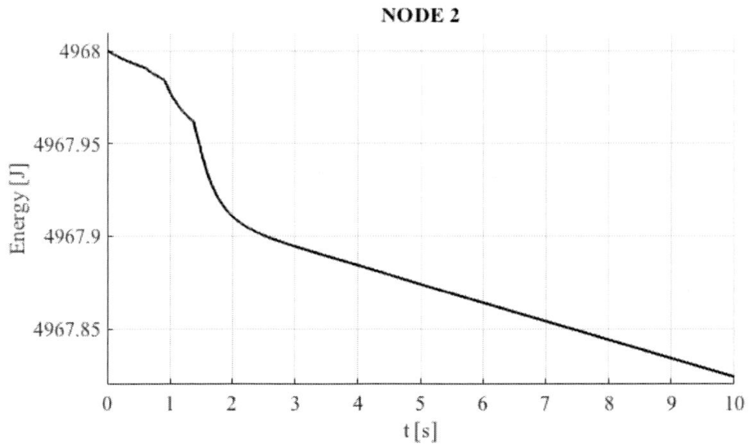

Figure 16. Battery consumption behavior in Node 2.

Figure 14 shows details of the use of the wireless network with their respective nodes in the MRI during the simulation.

Table 6 shows the processing of the network, of Node 1 and of Node 2 respectively with respect to sending or receiving data and communication test.

Figure 15 and Figure 16 shows the consumption behavior of the batteries during data transmission between the nodes.

The control performance was analyzed through ITAE and ISE as (Ogata, 2003).

ITAE

The value of the Integral of Time multiplied by the Error Absolute value - ITAE, is one of the most used performance indicators to evaluate control performance (Ogata, 2003).

It is based on the error of the controlled variable, for example, the difference between the current value of the process variable and the amplitude of the control signal.

The ITAE value is obtained according to (13) and is indicated to reduce the contribution of the performance value.

$$\text{ITAE} = \int_0^T t|e(t)|dt \tag{13}$$

The ITAE is indicated to reduce the contribution of large initial errors in the value of the integral of performance, as well as to emphasize the errors that happen later in the answer.

Figure 17 shows the performance analysis based on the error, in relation to Figure 8, referring to the difference between P_{ref} constant in Node 2 and P_m in the Node 1 sensor sent via WLAN wireless network to Node 2. Figure 18 presents the performance analysis based on the error, in relation to Figure 8, referring to the difference between P_{ref} constant in Node 2 and P_m in the Node 1 sensor sent via the ZigBee wireless network to Node 2.

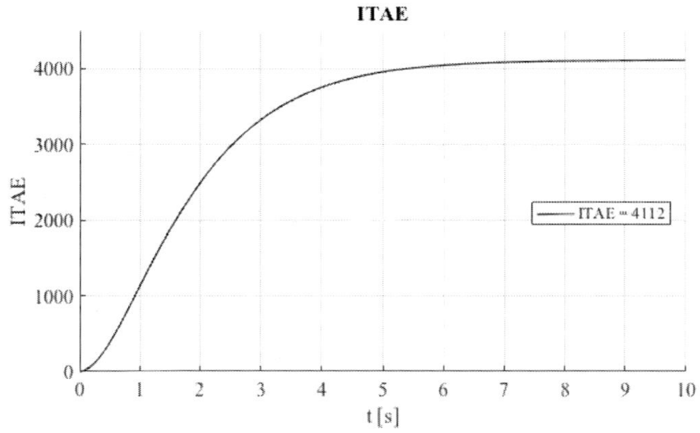

Figure 17. Performance analysis based on the error between P_{ref} and P_m through ITAE

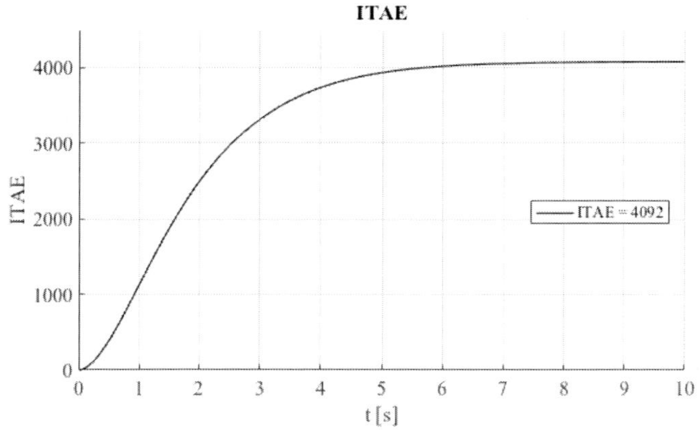

Figure 18. Performance analysis based on error between P_{ref} and P_m through ITAE

ISE

Another index of performance in the error, much used also is the Integral of the Square of the Error - ISE. This analysis is used to discriminate excessively superdamped systems from the sub-dampers, that is, it gives great weight to large errors and small weights to small errors, and because of this, it is not recommended to evaluate systems with only small errors (Ogata, 2003).

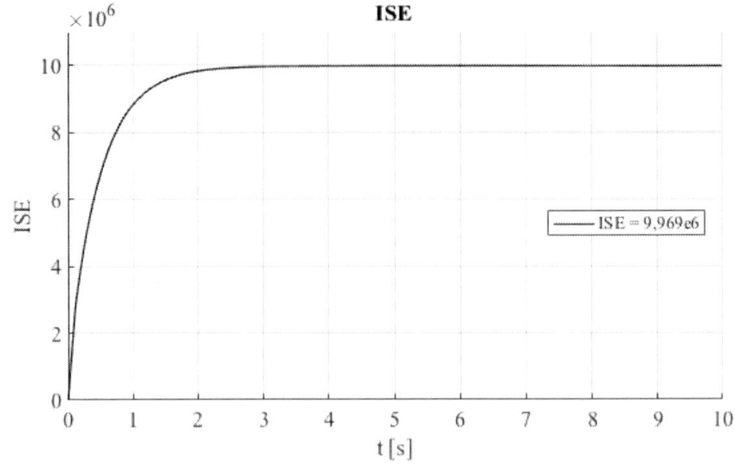

Figure 19. Performance analysis based on the error between P_{ref} and P_m through ITAE

The upper limit is a value obtained so that the integral has a steady value. The ISE value is obtained according to (14).

$$\text{ISE} = \int_0^T e^2\,(t)dt \tag{14}$$

Figure 19 shows the performance analysis based on the error, referring to the behavior of the use of the WLAN network in relation to the constant P_{ref} in Node 2 and the P_m in the sensor of Node 1 sent by wireless network to Node 2. While that the behavior of the processing of the use of the ZigBee network was identical even with respect to the value.

Through the presented data, it was possible to observe that the percentage of the use of the ZigBee network is much greater, when compared with the WLAN network. However, the waiting time of the ZigBee network is also much larger when compared to the WLAN network. Therefore, the time that the ZigBee network is off is 48.03% while the same as the WLAN network is 91.86%. Through these values, in this configuration it is possible to estimate that in the ZigBee network it would be possible to have at most two more DGs in the same circuit while in the Wi-Fi network it would be possible to estimate much more DG. This is due to the specific characteristics of each type of network, as shown in Table 1 and Table 5 respectively. In these configurations, it is possible to observe that the percentage of Node 1 processor usage in both nets presents very close values, as well as the Node 2 processor usage values.

The graphs of the performance analysis based on the error show that this configuration adopted for the smart MG simulation allows the use of the two types of network, that is, WLAN or ZigBee, since they are effective and viable in this analysis.

CONCLUSION

This chapter presents a study of the communication between distributed generation devices in a smart MG environment. The integrated simulation environment developed for smart MG incorporated the strengths of each tool in the same development environment as PLECS-Blockset software for electrical modeling, TrueTime-toolbox software for communication and real time analysis in Matlab-Simulink environment for the project of control. The flexibility and quantity of output parameters can be obtained with the integrated simulation scheme, since they represent a differential in the solution. For a MG scenario assuming a DG being the distributed controller, a node of the network was responsible for the control of the inverter and the measurements (Node 1), while the second node was responsible for the control of the active power (Node 2). The division of control between two devices allows the possibility to remotely access and control the active power generation using a smart MG in which suppliers and utilities would be connected and interact. By using the TrueTime-toolbox software, it was possible to model the communication network in the same environment in which the control and the electric circuit were modeled. In addition, some important variables on the exchange data, such as the length of time each message transmission was active or inactive and the battery discharge, were simulated and discussed. The overall performance of the communication was evaluated under different devices. For the MG studied in this study, both parameters had a small impact on the smart MG operation. It was possible to estimate that the ZigBee type network has a maximum of two DGs, whereas the WLAN network has a lot more, since, due to the specific characteristics of the network, it uses a percentage of network usage time, much smaller compared to the ZigBee network. Finally, this preliminary study is a starting point for a sort of other studies on the modeling of the communication network for smart MG. Additional studies may be conducted such as the influence of the distance between the nodes, the loss of packets during data transmission and the frequency of exchange of data on the control of the active power generation and the addition of other nodes in the network.

ACKNOWLEDGMENT

The authors thank the support for this work of the Group of Automation and Integrated Systems (GASI) of UNESP - Sorocaba/SP.

REFERENCES

[1] Vijayapriya, T. and Kothari, D. P. 2011. "Smart Grid: An Overview." In *Smart Grid and Renewable Energy*, edited by Tamilmaran Vijayapriya, and Dwarkadas P. Kothari, 305-311. India: Vindhya Institute of Technology & Science.

[2] El-hawary, M. E. 2014. "The Smart Grid—State-of-the-art and Future Trends." In *Electric Power Components and Systems*, edited by Mohamed E. El-hawary, 239-250. Canada: Dalhousie University.

[3] Mahmoud, M. S., Azher Hussain, S., and Abido, M. A. 2014. "Modeling and control of microgrid: An overview". *Journal of the Franklin Institute*. Accessed May 15, 2016.

[4] Guerrero, J. M., Blaabjerg, F., Zhelev, T., and Hemmes, K., Monmasson, E., Jemeï, S., Comech, P. M., Granadino, R., and Frau, I. J. 2010. "Distributed Generation: Toward a New Energy Paradigm." *IEEE Industrial Electronics Magazine*. Accessed Mar 18, 2016.

[5] IEEE, 2011. "IEEE Guide for Smart Grid Interoperability of Energy Technology and Information Technology Operation with the Electric Power System (EPS), End-Use Applications, and Loads." *IEEE Standard.* 2030-2011. Accessed Sep. 2015.

[6] Gungor, V.C., Sahin, D., Kocak, T., Ergut, S., Buccella, C., Cecati, C., and Hancke, G.P. 2011. "Smart Grid Technologies: Communication Technologies and Standards". *IEEE Transactions on Industrial Informatics.* Accessed Nov 2015.

[7] Souza, W. A., Marafão, F. P., Liberado, E. V., Diniz, I. S., and Serni, P. J. A. 2015. "Power Quality, Smart Meters and Additional Information from Different Power Terms." *IEEE Latin America Transactions.* Accessed Oct 20, 2015.

[8] Gungor, V.C., Sahin, D., Kocak, T., Ergut, S., Buccella, C., Cecati, C., and Hancke, G.P. 2013. "A Survey on Smart Grid Potential Applications and Communication Requirements". *IEEE Transactions on Industrial Informatics.* Accessed Feb 17, 2015.

[9] Sabbah, A.I., El-Mougy, A., and Ibnkahla, M. 2014. "A Survey of Networking Challenges and Routing Protocols in Smart Grids". *IEEE Transactions on Industrial Informatics.* Accessed Feb 19, 2015.

[10] NIST, 2014. *"NIST Framework and Roadmap for Smart Grid Interoperability Standards, Release 3.0."*, National Institute of Standards and Technology, Gaithersburg, MD, USA, 2014. http://dx.doi.org/10.6028/NIST.SP.1108r3. Available: http://www.nist.gov/smartgrid/framework3.cfm. Accessed: June, 2015.

[11] Shafiee, Q., Guerrero, J.M., and Vasquez, J.C. 2014. "Distributed Secondary Control for Islanded Microgrids — A Novel Approach." *IEEE Transactions on Power Electronics*. Accessed Jul 14, 2015.

[12] Cervin, A., Henriksson, D., and Ohlin, M. "TRUETUME 2.0 beta – Reference Manual." Department of Automatic Control, Lund University (June 2010). Accessed Nov 18, 2015. http://www.control.lth.se/attic/truetime/report-2.0-beta5.pdf.

[13] Cervin, A., Ohlin, M., and Henriksson, D., "Simulation of Networked Control Systems Using TrueTime." In: *3rd International Workshop*

on Networked Control Systems: Tolerant to Faults, Nancy, France. Jan 1, 2007.

[14] Godoy, Eduardo Paciência. *"Development of Networked Control Systems (NCS) for Applications in CAN-based Networks."* Written in Portuguese: (Desenvolvimento de Sistemas de Controle via Rede para Aplicações em Redes com Protocolo CAN). Doctoral Thesis, Universidade de São Paulo, São Carlos, 2011.

[15] Digi. 2016. *"XBee Wi-Fi. Embedded Wi-Fi modules provide ultra-low-power 802.11b/g/n communications in the flexible XBee hardware and software footprint".* Accessed Sep 14, 2016. https://www.digi.com/pdf/ds_xbeewifis6b.pdf.

[16] Thezolin, A. T., Ferreira, G. A. C. M., Júnior, J. A. L. V., and Bissoto, L. R. M. "*Rechargeable Batteries: Analysis of Its Advantages in Relation to Non-Rechargeable Batteries." 2010.* Written in Portuguese: (Pilhas Recarregáveis: Análise de suas Vantagens em Relação às Não Recarregáveis.) Accessed Feb 9, 2016. http://www.ib.unicamp.br/dep_biologia_animal/sites/www.ib. unicamp.br.site.dep_biologia_animal/files/EST.2010%20PILHAS% 20RECARREGÁVEIS.pdf.

[17] Digi. 2016. "XBee ZigBee. *Embedded ZigBee modules provide OEMs with a simple way to integrate mesh technology into their application".* Accessed Sep 21, 2016. https://www.digi.com/pdf/ds_xbee_zigbee.pdf.

[18] Ogata, Katsuhiko. 2003. *Modern Control Engineering*. Brasil: Sao Paulo. ISBN 978-85-87918-23-9.

[19] Plecsim – Electrical Engineering Software. 2002. *"PLECS – The Simulation Platform for Power Electronic Systems – User Manual – Version 4.0."* Accessed Feb 26, 2016. http://www.plexim.com/plecs.

In: Networked Control Systems
Editor: Eduardo Paciência Godoy
ISBN: 978-1-53613-105-5
© 2018 Nova Science Publishers, Inc.

Chapter 6

MULTI-RATE MODEL PREDICTIVE CONTROL FOR ENERGY EFFICIENCY IN WIRELESS NETWORKED CONTROL SYSTEMS

Felipe Fakir * ***and Eduardo Paciência Godoy***
São Paulo State University (UNESP), Sorocaba, SP, Brazil

ABSTRACT

Wireless communication technology has become a fundamental part of the process industries, where the use of wireless transmitters for monitoring and control is a reality. The architecture of Wireless Networked Control Systems (WNCSs) has advantages over point-to-point and wired networks architectures. However, the evolution of this technology has introduced new challenges to the implementation of the closed-loop control with a wireless instrument. A critical factor related to implementation of WNCSs is the energy source of these transmitters, which have limited lifetime dependent on the amount of access and data transmitted. This chapter presents the study and development of a multi-

* Corresponding Author Email: zafakir@yahoo.com.br.

rate model predictive control (MPC) as an alternative to improve energy efficiency in industrial applications of WNCSs. The proposed strategy uses less process variables data as it does not requires frequently updates of the process variables transmitted by the wireless instruments. The MPC uses the internal submodel of the process variables to estimate the process signals when they are not transmitted. Thus, a decrease in the frequency of data transmission on the wireless network can be obtained and consequently a reduction of energy consumption of wireless devices. Simulation results of a multivariable WNCS of coupled tanks proves that the multi-rate MPC provides energy economy as it is effective in saving the energy expenditure of the wireless transmitter's battery. An energy efficiency analysis of the WNCS is presented by studying the operating limits of the multi-rate MPC considering the compromise relationship between the sampling period of the wireless devices and the control performance of the WNCS.

Keywords: wireless networked control systems, energy efficiency, model predictive control, multi-rate control

INTRODUCTION

Several benefits can be observed within the wireless technology when compared to traditional process control systems that uses wired instruments. As an example of its advantages, it is mentioned the increase of mobility, flexibility and ease of installation, absence of specific connectors, reduction of the costs of the project in the construction stage and avoidance of unforeseen problems such as open-loop circuits [1, 2]. Due to these benefits, wireless technology is presented as an attractive option for the process industries when used in conjunction with distributed control systems [3, 4].

With the standardization of the industrial protocol WirelessHART in 2010 by IEC 62591 [5], the process industries are in recent modernization to adapt to the new era of these electronic devices, in this way, several applications and studies has been developed. One of the areas with great development potential is wireless networked control systems, also known as WNCS [6].

Currently, in the process industries, the application of closed-loop control using wireless transmitters is practically non-existent when compared to the conventional control strategies in which wired transmitters are used. This is due to some challenges encountered in the industry such as severe operating environments and the presence of noise [7]. In addition, when the WNCS is applied, the following points can be observed that contribute to the lack of reliability in the industrial sector [8, 9]:

- Loss of packets in data transmission;
- Delays in data transmission at the control network layer;
- Low autonomy level of battery life when the wireless nodes are operating at fast sampling rates;

The low demand for this type of application is also related to the conservative nature of the process industry. Today, these wireless transmitters are typically applied in the field of process variables monitoring. However, the growing interest in the industrial sector in using these instruments for closed-loop control and in WNCS has recently been noted [5].

Control techniques applied to the wireless network are being developed and exploited with the purpose of enabling the operation and closed-loop control of industrial processes in which the wireless instruments perform the key function of measurement of the control variable [2]. One of the great challenges in the usage of wireless instruments in closed-loop control applications is based on the process variable update period in relation to the battery consumption of the wireless transmitter [10].

The wireless transmitters have great limitations regarding the autonomy of the battery level, therefore the energy efficiency of the sensors must be considered. Typically, these transmitters, for the most part, must remain operating in sleep mode, where transmitting or receiving data is not allowed. Therefore, this requirement does not favor the use of this technology applied to closed-loop control, since the regulatory requirements may require high demand in the information transmission rate using these transmitters [2]. The relationship between the process variable update rate of a wireless transmitter is inversely proportional to the battery level autonomy, so the faster the transmitter data is sampled and transmitted, the shorter the battery life [11].

Energy efficiency can be defined as the possibility of improving the usage of the energy consumption associated with an activity, thus contributing to the reduction of expenditure with this input. In a simplified way, this means performing the same activity using less energy. From the WNCS point of view, energy efficiency seeks to reduce the power consumption of wireless devices, so that the lifetime of the battery is extended without the control performance of the WNCS being significantly impaired [12].

When it comes to reducing energy expenditure, it is known that data transmission in the wireless network consumes an important amount of energy in a wireless device and is the main consumer of energy in a large class of applications [11]. The challenge, therefore, is to extend the useful life of the batteries that feed these wireless devices to the maximum without degrading the control and stability performance of the WNCS. Thus, a common approach to energy savings in WNCS has been the reduction of wireless transmissions [9].

To improve the energy efficiency of the WNCS and to overcome the various challenges encountered in the use of wireless transmitters applied in the field of control of industrial processes, this chapter presents the study of a multi-rate model predictive control strategy. In addition, it is of interest in this chapter to present, contribute and reinforce the applicability of wireless transmitters in the closed-loop control for the process industry.

Multi-Rate Model Predictive Control

PID control is the most used in the process industry, however, when the process dynamics is complex and has some characteristics such as nonlinearities, coupled dynamics among others, different control strategies may be considered to obtain better control performance. Model predictive control (MPC) can replace conventional control strategies and solve control problems resulting in a better system performance. In addition, the MPC can cover systems with more than one variable, respecting the constraints and limits of the process, as well as being able to operate as an optimizer, minimizing the cost functions associated with the control problem.

The MPC uses an algorithm in which the process model is used to predict the future state of the system and thereby predict the future system response and the control actions necessary to minimize the desired cost function [13]. These characteristics give the MPC the anticipatory feature mainly because it uses the future estimates to calculate the current control signal [14]. Logically, the MPC can also be called optimal control, because it is possible to consider a desired cost function to obtain a process optimization, considering the limits and restriction of the controlled process.

Figure 1 shows the multi-rate MPC in operation uses the process model, which is the combination of several internal submodels with different process dynamics and / or sampling periods. During the execution, the process submodel with the fastest process dynamics has its actual measurement used at each multi-rate MPC execution cycle. The submodels with slower process dynamics should have their actual values used in the multi-rate MPC only at the moment their measurement coincides with the controller execution cycle.

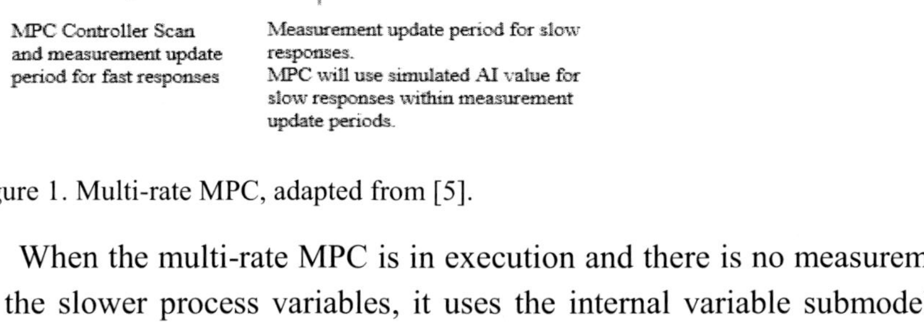

Figure 1. Multi-rate MPC, adapted from [5].

When the multi-rate MPC is in execution and there is no measurement of the slower process variables, it uses the internal variable submodel to estimate its value. By definition, the control system is operating in a multi-rate mode when the control strategy is executed and the signals are sent to the actuators N times faster than the sensors sampling. When the reading cycle of the variable with the fastest dynamic matches the one with the slowest process dynamic, the actual measurement values can be used to update the multi-rate MPC internal submodels, in which case the model optimization is enabled during the control execution [5].

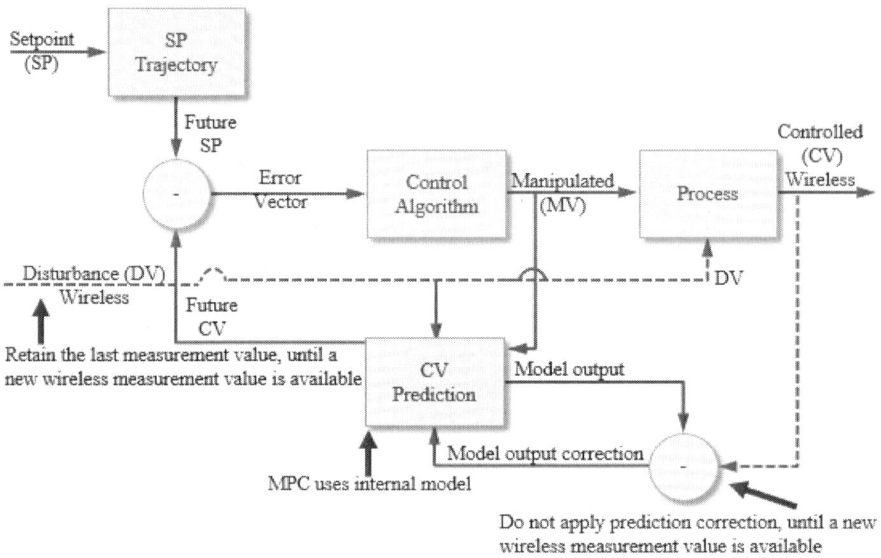

Figure 2. Basic structure of a wireless MPC controller, adapted from [5].

In WNCSs, the multi-rate MPC strategy can be used to optimize the energy consumption by predicting future states and using them as inputs to the control algorithm. There is no need for the controller to access the current real data through the wireless network [15].

Changes in the original structure of the MPC consist of not correcting the internal process model used by the controller if the actual data is not available. In addition, the controller uses the actual data to control and only updates the internal model if the optimization key is enabled and when the actual data are available. Finally, the controller uses the internal model to calculate virtual measurements of the process to be used by the control algorithm and calculates the value of the estimated output, as can be observed in Figure 2.

MATERIALS AND METHODS

The control strategy presented in this chapter is the multi-rate MPC applied to a control system via wireless network (WNCS) in the field of industrial process control in which the sampling rate of the sensors exceeds the limit of seconds. The purpose of the strategy is not to improve the control performance of the WNCS, but rather to increase the energy efficiency of the system by reducing the amount of data transmissions of the sensors through the wireless network, in view of the performance of the control system within an acceptable range. Figure 3 depicts the general block diagram of the proposed system architecture and its components. The principle of operation is based on the transmission of signals from the sensors through a wireless network combined with virtual signals, predicted by the MPC, the signals are used in the calculation of the control signals and finally these are transmitted to the actuators. As the controller makes use of virtual data, it is possible to increase the sampling time of the sensors by reducing the volume of data transmission in the wireless network and consequently increasing the energy efficiency of the system.

The selected process plant AMIRA DTS200 system is presented by Figure 4, which is composed of three coupled tanks, sensors and actuators. Details about the system can be found in [17]. This system was chosen as the process plant to be controlled by the wireless control system using a multi-rate MPC.

The subsequent step was the development of the mathematical modeling of the process plant and the development of the process model block with its respective system dynamics. After the creation of the dynamic system model block, the multi-rate MPC was developed for a wireless network control system (WNCS).

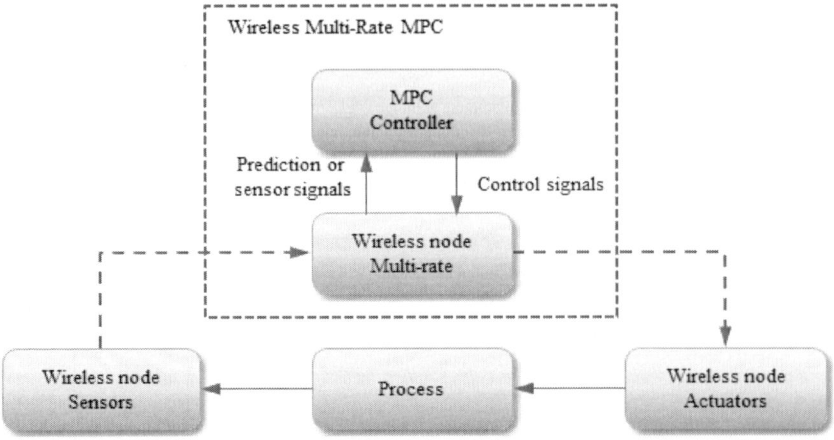

Figure 3. General diagram of the WNCS architecture.

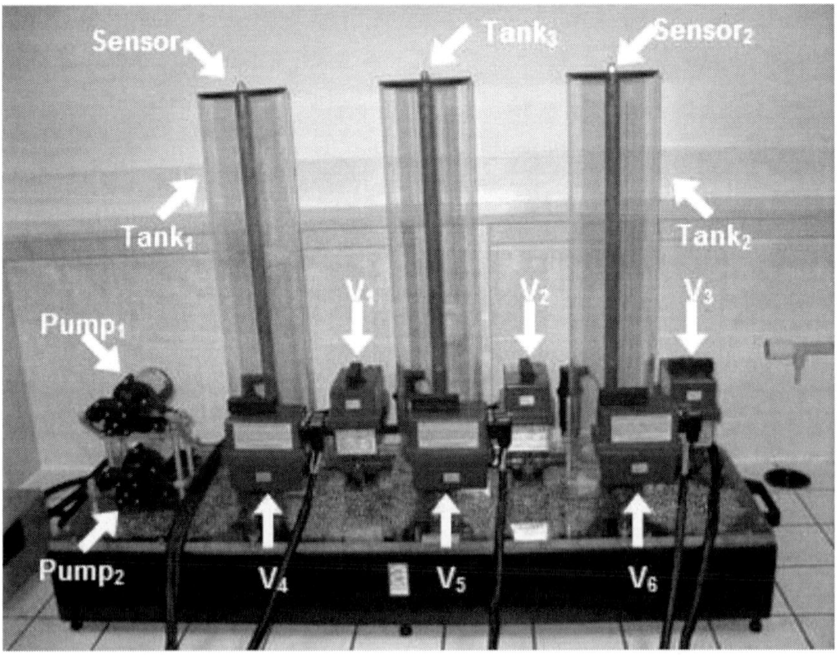

Figure 4. Three-tank process plant structure, adapted from [18].

The WNCS was integrated to the model block of the process, several simulations were performed and the results were collected. Closed loop control performance indexes were generated for several simulations. The analysis used the value of the ErrorRMS and normalized ITAE (based on the worst performance) and the graphics for performance evaluation were generated for the energy efficiency analysis of the proposed simulated system.

Development of the Mathematical Model

The mathematical modeling of the AMIRA DTS200 plant uses the variables presented in Figure 5.

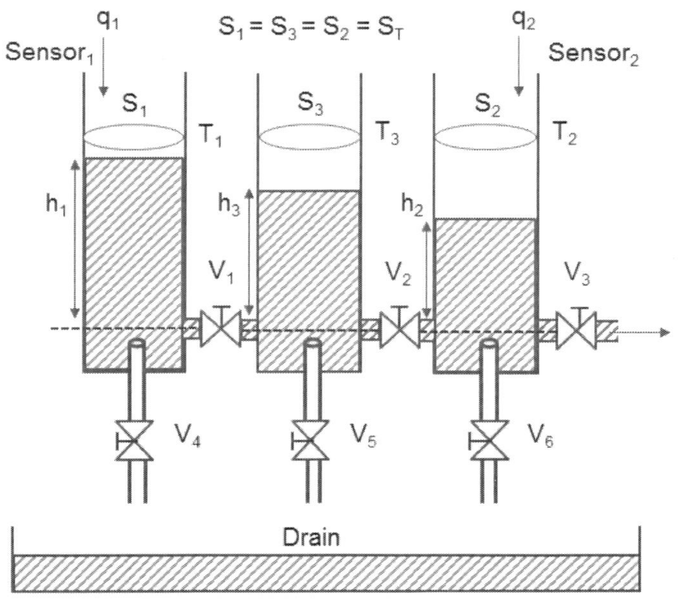

Figure 5. AMIRA DTS200 three-tank process plant, adapted from [18].

The flow rates in each of the valves can be calculated by the Torricelli rule described by (1), where az_n is the flow coefficient at the valve outlet n, a dimensionless value that can vary from 0 to 1, Δh represents the difference between two levels, g being the acceleration of gravity and finally the area of the maximum cross-section of the valve is S_n.

$$q_n = az_n\, S_n\, sign\,(\Delta h)\,(2g\,|\Delta h|)^{\frac{1}{2}} \tag{1}$$

The parameters h_1, h_2 e h_3 representing the height of the liquid level in the tanks 1, 2 and 3, respectively. The parameters $q_{V1}, q_{V2}, q_{V3}, q_{V4}, q_{V5}$ and q_{V6} representing the liquid flow in the valves V_1, V_2, V_3, V_4, V_5 and V_6 respectively. The parameter S_T as the cross-sectional area of the tanks, and finally, q_1 and q_2 representing the volumetric flow of pumps 1 and 2. These parameters were used to generate the mathematical model of the system.

It is possible to describe the flow of the system using the concept presented in (2), in which the variable A represents the cross-sectional area, $\frac{dh}{dt}$ the rate of change of the level by time and $\sum q$ the sum of all the flows in the tank under analysis.

$$A \cdot \frac{dh}{dt} = \sum q \tag{2}$$

Applying the concept of (2) to Tanks 1, 2 and 3 respectively, we obtain (3), (4) and (5) which governs the dynamic behavior of the system in each tank individually.

$$S_T \cdot \frac{dh_1}{dt} = q_1 - q_{V1} - q_{V4} \tag{3}$$

$$S_T \cdot \frac{dh_3}{dt} = q_{V1} + q_{V2} - q_{V5} \tag{4}$$

$$S_T \cdot \frac{dh_2}{dt} = q_2 - q_{V2} - q_{V3} - q_{V6} \tag{5}$$

Substituting the variables $q_{V1}, q_{V2}, q_{V3}, q_{V4}, q_{V5}$ and q_{V6} that represent the liquid flow in the valves by (1), rule of Torricelli, we obtain the general mathematical model of the plant to be controlled.

DEVELOPMENT OF MULTI-RATE MPC FOR WNCS

A MPC controller has the ability to handle multiple input and output (MIMO) systems. In addition, it is also usually considered in the control of nonlinear systems the possibility of automatic adjustments in the internal model used in the prediction of the future state and in the control signals.

Figure 6 presents the general concept in the adaptation of conventional MPC to MPC multi-rate. In a conventional control system, the execution of the control block occurs at each *T* s, so the sensor reading, the controller calculation and the plant actuation must occur in the same cycle, that is, at the same rate of *T* s.

In multi-rate MPC, it is expected that the control block and transmission of the control signals to the actuators will be executed at each *T* s. However, transmitting the values of the sensors to the control block can occur in multiple sample periods of *T* s. It happens respecting the premise that the process variable with the fastest dynamic is transmitted at the same frequency as the execution period of the control block. If the value of the process variable is not available in each execution cycle, the control block uses the internal sub-model to calculate the control signal.

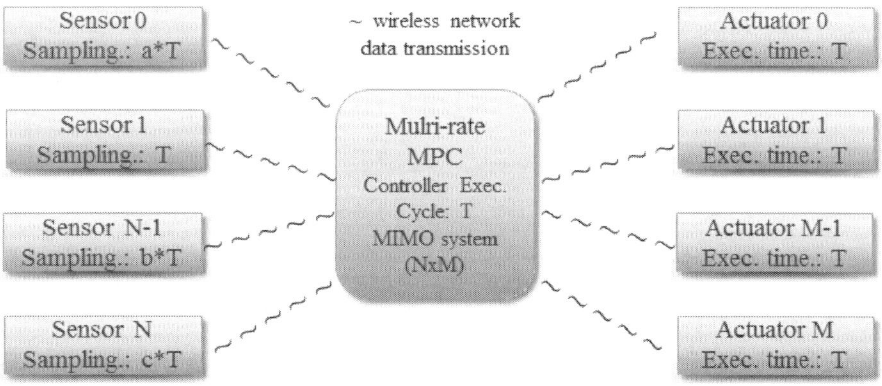

Figure 6. Multi-rate MPC for WNCS.

The transmission period of the data by the sensors operating at multi-rate must respect the common rule of multiples. Thus, the constants a, b and c shown by Figure 6 must have a common multiple divisor, so that in the period in which the block the actual reading of all sensors, it is possible to update the parameters of the internal process model with the actual data.

The multi-rate MPC developed for applications in WNCS is represented by Figure 7, which presents a block diagram with all the components involved in the system. This diagram was built in the Matlab/Simulink environment along with the TrueTime toolbox and the MPC toolbox.

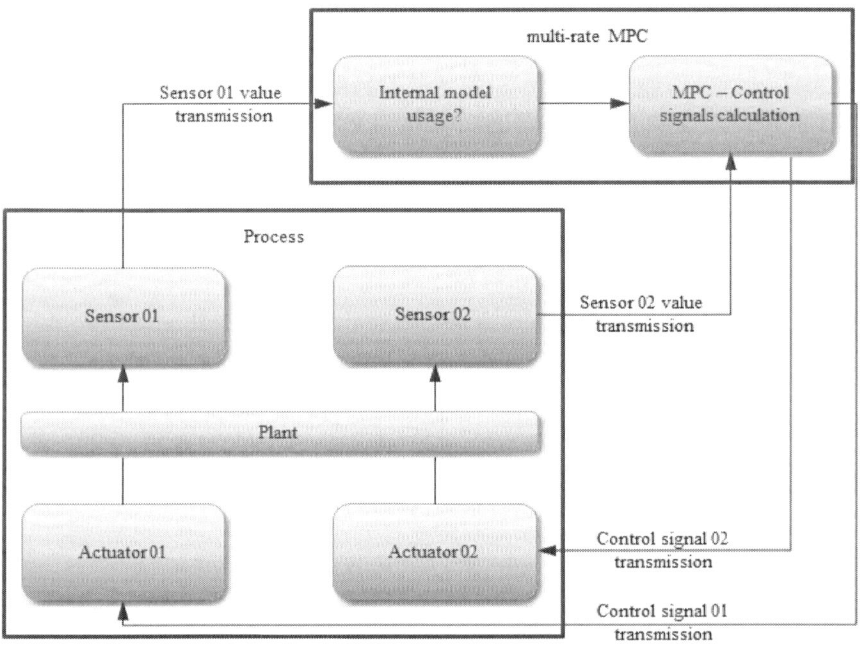

Figure 7. Multi-rate MPC control block diagram for WNCS.

The development platform for the simulation environment chosen was Matlab/Simulink and TrueTime [16] and MPC toolbox were used. TrueTime is a toolbox for the Matlab / Simulink platform used to facilitate the simulations of tasks such as controllers, transmission networks and continuous dynamic systems like WNCS with real-time requirements. The toolbox contains the following blocks:

- Wireless Network (Wireless network setup);
- Kernel (Converters A/D, D/A and logic execution);
- Battery (Measures the device's battery consumption);

RESULTS AND ANALYSIS

Multi-Rate MPC Performance Analysis

For the performance analysis of the proposed multi-rate MPC, simulations were performed and the response curves and values of the performance indexes were collected. The experiments consisted in observing the system response with different sampling rates in the absence and presence of disturbances. The profile reference was defined to avoid saturation of the actuators, slope ramps for T_1 and constant reference signals on T_2 were defined.

The results were analyzed numerically by calculating the value of the ErrorRMS and normalized ITAE. The ErrorRMS is given by the gross value and the normalized ITAE is given by dividing the gross value obtained by the highest value in all the experiments. In that manner, the worst performance is represented by a unitary value of the normalized ITAE and the best performance is the one that approaches zero value.

Simulations performed by keeping the multi-rate MPC execution period and the sensor$_2$ (T_{S2}) sampling period at the lowest available value for this set, which is 1 s. The value of the sampling period of sensor$_1$ (T_{S1}) was varied in 1, 4, 8, 16 and 32 s resulting in the graphics found in Figure 8. Since multi-rate MPC operates in execution cycles faster than the rate of sensor$_1$, the controller makes use of the predicted values through the internal submodel of the process variable, which has uncertainties regarding the actual model of the plant. This occurs because in several executions cycles the actual value of sensor$_1$ is not available for calculation of the control signal. In general, the analyzes performed on this multi-rate MPC strategy point out that the increase of the sampling period of $sensor_1$ influences the control mesh performance in a WNCS, resulting in higher values of ErrorRMS and normalized ITAE.

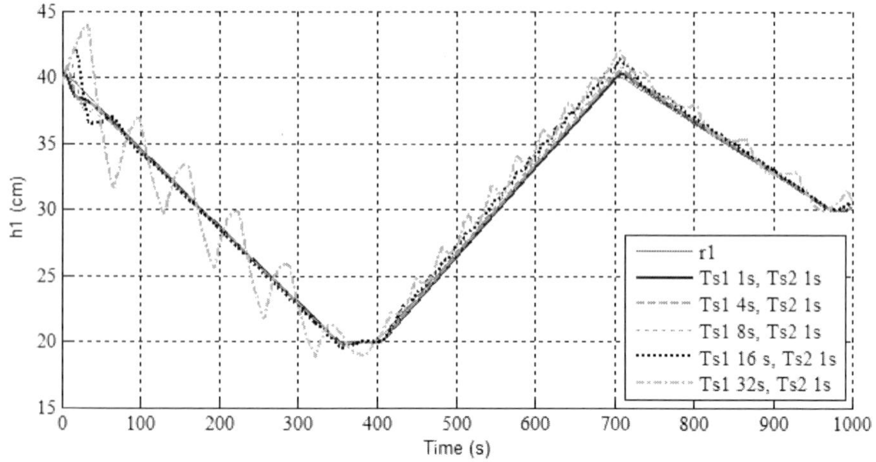

Figure 8. Data from h1 in the experiment with TS2 at 1 s.

The curves referring to the measurement of the level in the $tank_1$ with respect to the set point profile (r_1), found in Figure 9, show that by increasing the sampling period of $sensor_1$, the curves become tenuously more oscillatory, resulting in increment of the values of the ErrorRMS and normalized ITAE. Note that when performing the simulation with the sampling period from $sensor_1$ to 32 s, the system may become oscillatory at some intervals, demonstrating that it would be impracticable to work with the multi-rate MPC with this $sensor_1$ sampling.

To verify the robustness of the controller, it was proposed to operate the WNCS in the presence of disturbances. The experiment with disturbance was performed using the multi-rate MPC execution period and the sampling period of $sensor_2$ (T_{S2}) in the value of 1 s. For the sampling period of $sensor_1$ (T_{S1}), the values were varied from 1, 4, 8, 16 and 32 s.

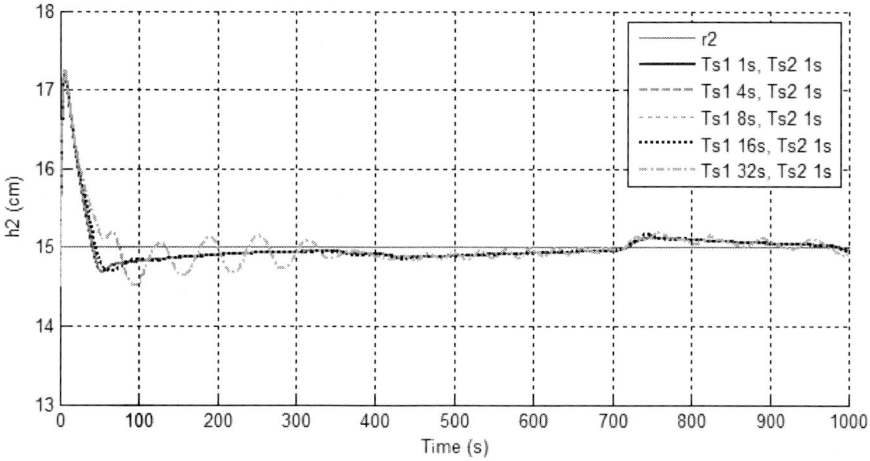

Figure 9. Data from h2 in the experiment with TS2 at 1 s.

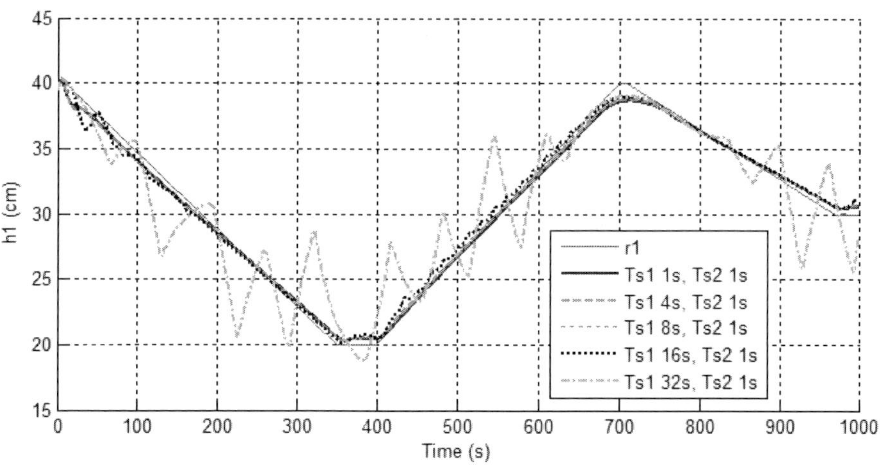

Figure 10. Data from h1 in the experiment with sinusoidal disturbance with TS2 at 1 s.

The simulations were performed in the presence of a sinusoidal disturbance of 30 ml/s amplitude with oscillation frequency of 0.01 Hz in the feed rate in $tank_1$. For this experiment, the graphical results of the response curves of h_1 and h_2 are shown in Figure 10.

Figure 10 shows that the multi-rate MPC controller obtained a good response to the sinusoidal disturbance inserted in the system, excluding the case where the T_{S1} value is 32 s. It is noteworthy that in executing the simulation with these sample period values, the system proved to be robust, making it possible to apply the WNCS under these conditions.

Trend Analysis of the Performances Indexes

For the analysis of the system operational limits, several simulations were carried out. For this, the mapping of the possible scenarios covering all the possible sampling periods of the sensors was done and the performance indexes were collected and used in this section.

The analysis of the trend of the ErrorRMS and normalized ITAE indexes as a function of the sensor sampling period in a WNCS was used to present the operational limits of the system. Figure 11 shows the performance index curves of the MPC controller relative to the level control of the T_1 as a function of the variation of the parameter T_{S1}, keeping the value T_{S2} (1 s) constant.

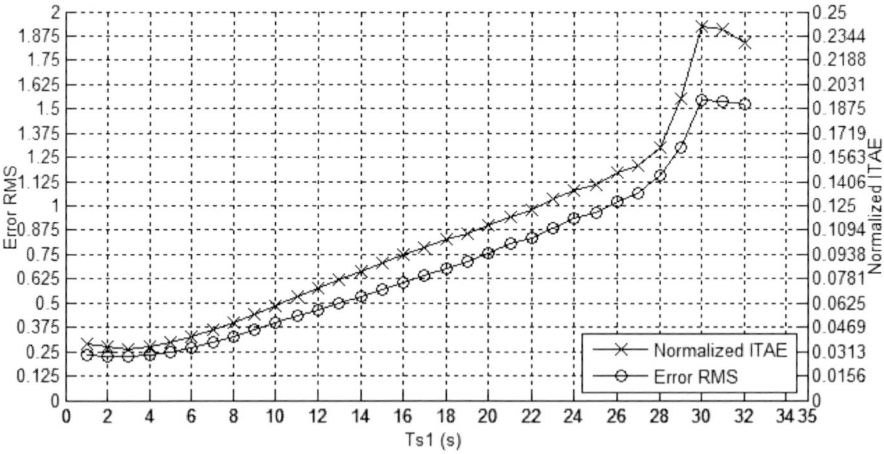

Figure 11. Trend of the ErrorRMS and normalized ITAE indexes (h1 vs TS1).

The curves of Figure 11 show four regions with different behaviors as in Table 1.

Table 1. Performance Indexes based on T_{S1} (s)

T_{S1} (s) [Range]	Index Variation
1 – 5	Minimum
5 – 27	Linear
27 – 30	Exponential
30 – 32	Non-Uniform

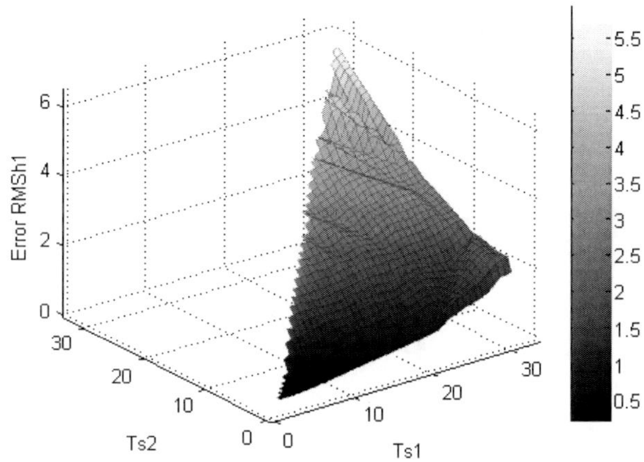

Figure 12. Trend of the ErrorRMS index for h1 vs TS1 and TS2.

The curves shown in Figure 11 and the results presented in Table 1 demonstrate different behaviors. By increasing T_{S1}, it can be seen that the curves of the ErrorRMS and normalized ITAE indexes have the same trend. In a first moment, there is a minimum variation in the indexes, soon after, a linear behavior is noticed and then exponentially, finally the system has non-uniform behavior due to the oscillations of the process variable around the reference curve. The results suggest that the acceptable limits of the ErrorRMS and normalized ITAE indexes under these conditions should be lower than 1.067 or 0.151 respectively, because, when exceeding these thresholds, low controller performance is found and the system starts operating in the region of exponential growth or not uniform. Keeping the indexes within the proposed limits, the system operates with adequate performance.

In general, the behavior of the performance indices was investigated according to the sampling period of the sensors. Figure 12 shows the curve of the ErrorRMS index of the controller for the T_I as a function of the variations of the T_{S1} and T_{S2} parameters. The normalized ITAE index curve follows the same trend as the ErrorRMS index as shown in Figure 12, resulting in a similar surface.

Controller and WNCS Energy Efficiency

The energy efficiency analysis was based on simulations carried out to investigate the battery life of the wireless devices that transmit data through the WNCS wireless network. Among all simulation results, only those with the acceptable limit conditions of the normalized ErrorRMS and ITAE indexes (less than 1,067 or 0,151 respectively) were selected, and the consumption of each of the elements transmitting data over the wireless network of WNCS.

For the measurement of energy consumption, the information provided through the TrueTime Battery block connected to the sensors and the multi-rate MPC was used. Table 2 shows the energy consumption in mW (J/s) and the duration of each of the functions performed by the sensor using the XBee S2B module (Series 2 Model B) when the sleep, idle, measurement or transmission functions of 2 bytes of data are executed [19, 20, 21].

Table 2. Sensor Consumption based on XBee S2B Model

Function	Consumption (mW)	Duration (ms)
Sleep	26,4	Variable
Idle	72,6	Variable
Measurement	82,5	1
Data Transmission	693,0	18

Table 3 shows the energy consumption in mW (J/s) and the duration of each of the functions performed by the controller using the XBee S2B module when the idle, processing and 2-byte data transmission functions are performed [19, 20, 21].

Table 3. Controller Consumption based on XBee S2B Model

Function	Consumption (mW)	Duration (ms)
Algorithm Execution	72,6	Variable
Data Transmission	99,0	5

The consumption of the actuators was not taken into account, since they do not transmit data through the wireless network. It is worth emphasizing that the focus of this work seeks to increase the energy efficiency in WNCS by reducing the energy consumption of the network elements, so reducing the amount of information flow through the reduction of data transmission over the wireless network, consequently the system becomes more efficient considering the energy question [21].

Table 4 shows the energy expenditure of the network devices (Multi-rate MPC sensors and controller) and the overall energy consumption of the WNCS, focusing on the devices that transmit data. The energy consumption values for the sensors were obtained based on the data provided by Table 2, in these simulations were considered the functions of measurement, transmission and sleep, the energy consumption of the controller was obtained based on the data provided by Table 3 and the overall consumption of WNCS was calculated in each case.

The results were analyzed numerically by calculating the efficiency of the WNCS, calculated as the percentage ratio between the WNCS consumption in the scenario under analysis and the highest energy consumption value of the system, in which case the network elements operate at the rate of sampling equivalent to 1 s. Therefore, the highest energy efficiency obtained for the WNCS is the one with the highest value in %, and the efficiency of the WNCS is presented in the last column on Table 4.

Table 4. WNCS Consumption based on the Indexes Filter (1000 s of Simulation)

T_{S1} (s)	T_{S2} (s)	S_1 (J)	S_2 (J)	MPC (J)	WNCS (J)	WNCS (%)
1	1	38,795	38,795	84,579	162,168	0,000
27	1	26,871	38,795	84,579	150,245	7,352
2	2	32,597	32,597	78,589	143,784	11,336
22	2	26,970	32,597	78,590	138,157	14,806
3	3	30,540	30,540	76,601	137,681	15,100
21	3	26,995	30,540	76,601	134,136	17,286
4	4	29,499	29,499	75,595	134,592	17,005
20	4	27,020	29,499	75,595	132,113	18,533
5	5	28,879	28,879	74,996	132,754	18,138
10	5	27,639	28,879	74,996	131,514	18,903
6	6	28,470	28,470	74,601	131,540	18,886

Results of the simulations confirm that decreasing the frequency of sampling/transmission of data in the wireless network by the network devices can increase the energy efficiency of the WNCS within the acceptable ranges of control performance. In a practical case, the best sampling strategy for the multi-rate MPC would be obtained with $sensor_2$ and the multi-rate MPC execution cycle by sampling and transmitting the data every 6 s. Defining the same sampling period, 6 s, for both wireless sensors would reduce maintenance and control system stops to replace the sensor batteries (avoiding multiple interventions in the process to replace the batteries alternately).

Figure 13 shows the energy efficiency curve of the WNCS as a function of the variations of the parameters T_{S1} and T_{S2} using the data in Table 4. Increasing the sampling periods of the system's T_{S1} and T_{S2} sensors results in higher energy efficiency.

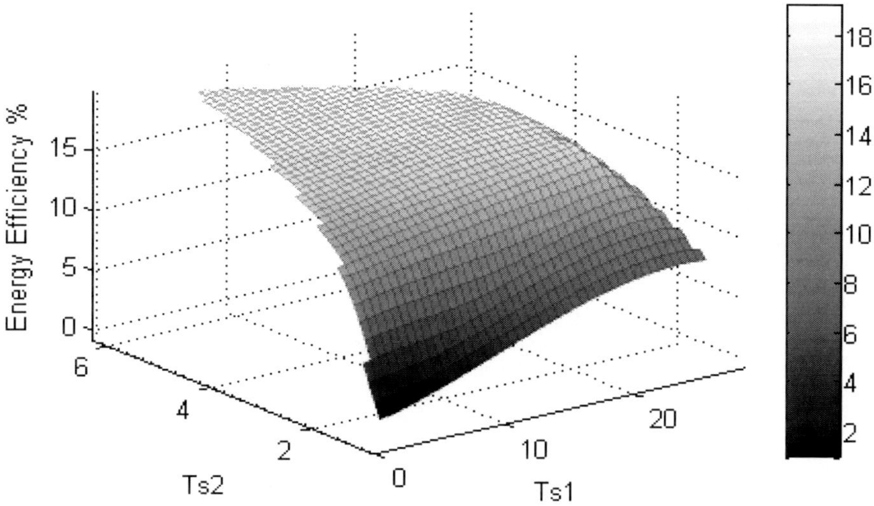

Figure 13. 3D trend of energy efficiency as a function of TS1 and TS2.

Table 5. WNCS Consumption with Sensors using Sleep Mode (1000 s of Simulation)

T_{S1} (s)	T_{S2} (s)	S_1 (J)	S_2 (J)	MPC (J)	WNCS (J)	WNCS (%)
6	6	74,523	74,523	74,601	223,647	-37,911

Table 5 shows the energy expenditure of the network devices and the overall energy consumption of the WNCS, focusing on devices that transmit data by changing only the function from sleep mode to idle.

The result of Table 5 shows that there is a large difference between the use of the sleep and idle modes for the WNCS sensors. The WNCS becomes less efficient in terms of energy since when the sensor operates with idle and sampling intervals at 6s more energy is spent when compared to sleep mode operation and sampling 1s with negative energy efficiency (-37,911%). Therefore, it is confirmed that using the sleep function would be possible to increase the energy efficiency of the WNCS.

Conclusion

In this chapter, a model-based multi-rate predictive controller (MPC) was developed for WNCS to control a non-linear multivariable system with the objective of increasing the energy efficiency of the WNCS. The results of the experiments made it possible to analyze the performance of the multi-rate MPC and to observe the influence of the increase of the sample period of the wireless devices on the WNCS. In addition, it was possible to determine the system operating sampling rate in which WNCS has an excellent compromise between control quality and energy efficiency.

In the simulation environment, it was possible to identify the operational limits of the system, to which the ErrorRMS and normalized ITAE should be less than 1,067 or 0,151 respectively. Considering the limit values of established performance indexes, the optimal compromise values were found between the sensor sampling period and the energy efficiency of the system, when the sampling rates of the system elements operates at 6 s, the best result energy efficiency is achieved. This rate when compared to conventional 1 s for the system, demonstrates that the proposed WNCS has the ability to operate reducing system power consumption by up to 18.886% while maintaining the quality of control and stability.

Through all the collected results, it is verified that multi-rate MPC applied to the WNCS presented the ability to maintain the same stable and with adequate control performance with high sampling periods and in the presence of disturbances with high energy efficiency. It is pointed out that multi-rate MPC applied to a WNCS is feasible and increases the energy efficiency of the system with the application of this technique.

Acknowledgment

Research supported by grant 2015/01303-2, São Paulo Research Foundation (FAPESP).

REFERENCES

[1] Song, Jianping., Song, Han., Mok, Aloysius K., Chen, Deji., Lucas, Mike., Nixon, Mark., and Wally Pratt. 2008. "WirelessHART: Applying Wireless Technology in Real-Time Industrial Process Control." *Proceedings of the 14th IEEE Real-Time and Embedded Technology and Applications Symposium (RTAS)*, St. Louis, pp. 377-386.

[2] Willig, Andreas. 2008. "Recent and Emerging Topics in Wireless Industrial Communications: A Selection." *IEEE Transactions on Industrial Informatics*, v. 4, n. 2, pp. 102-124.

[3] Willig, Andreas, Kirsten, Matheus, and Adam Wolisz. 2005. "Wireless technology in industrial networks." *Proceedings of the IEEE*, v. 93, n. 6, pp. 1130–1151.

[4] Moyne, James R., and Dawn M. Tilbury. 2007. "The emergence of industrial control networks for manufacturing control, diagnostics, and safety data." *Proceedings of the IEEE*, v. 95, n. 1, p. 29–47.

[5] Blevins, Terry., Chen, Deji., Nixon, Mark., and Willy K. Wojsznis. 2015. *Wireless Control Foundations: Continuous and Discrete Control for the Process Industry*. USA: International Society of Automation.

[6] Fischione, Carlo., P. Park., P. Di Marco, Piergiuseppe., and Johansson, Karl H., 2011. "Design Principles of Wireless Sensor Networks Protocols for Control Applications." *S. K. Mazumder (Ed.). Wireless Networking Based Control*. New York: Springer, 2011. Chap. 9, pp. 203-238.

[7] Song, Han., X. Zhu, Mok., Aloysius K., Deji, Chen., and Mark Nixon. 2011. "Reliable and Real-Time Communication in Industrial Wireless Mesh Networks." *IEEE Real-Time Technology and Applications Symposium*, Chicago, IL, 2011, pp. 3-12.

[8] Baillieul, John., and Panos J, Antsaklis. 2007. "Control and Communication Challenges in Networked Real-Time Systems." *Proceedings of the IEEE*, v. 95, n. 1, pp. 9-28.

[9] Araújo, José., Mazo, Manuel., Anta, Adolfo., Tabuada, Paulo., and Karl H. Johansson. 2014. "System Architectures, Protocols and Algorithms for Aperiodic Wireless Control Systems." *IEEE Transactions on Industrial Informatics*, v. 10, n. 1, pp. 175-184.

[10] Akerberg, John., Gidlund, Mikael., and Mats Björkman. 2011. "Future research challenges in wireless sensor and actuator networks targeting industrial automation." *Proceedings of the 9th IEEE International Conference on Industrial Informatics (INDIN)*, Lisboa. 2011, pp. 410-415.

[11] Wen, Rongrong., Mareels, Iven., and Brian Krongold. 2007. "Optimal power analysis for network lifetime balance in wireless sensor networks." *Proceedings of the 2007 International Symposium on Communications and Information Technologies*, Sydney, NSW, 2007, pp. 1161-1165.

[12] Mansano, Raul K. 2016. *"Multi-rate Adaptive Control for Energy Efficiency in Wireless Networked Control Systems* [In Portuguese]." Master Dissertation on Electrical Engineering, São Paulo State University (Unesp), Bauru, Brazil.

[13] Darby, Mark L., and Michael Nikolaou. 2012. "MPC: Current practice and challenges." *Control Engineering Practice*, v. 20, n. 4, pp. 328-342.

[14] Santana, Eudemario Souza. 2007. "*Model Predictive Algorithm applied to the Induction Motor Velocity Control* [In Portuguese]." PhD Thesis in Electrical Engineering, State University of Campinas (Unicamp), Brazil.

[15] Li, Huiping., and Yang Shi. 2013. "Output feedback predictive control for constrained linear systems with intermittent measurements." *Systems & Control Letters*, v. 62, n. 4, pp. 345-354,

[16] Cervin, Anton., Henriksson, Dan., and Martin Ohlin. 2016. *"TrueTime 2.0 – Reference Manual."* Accessed Jan 22, 2017. Available in: <http://www.control.lth.se/truetime/>.

[17] Amira. 2002. *"DTS200 Laboratory Setup Three-Tank-System."* Accessed Jan 22, 2017. Available in: <http://www.utm.mx/~jvasquez/tanques_lab.pdf>.

[18] Chalupa, Petr., Novák, Jakub., and Vladimír Bobál. 2011. "Detailed Simulink Model of Real Time Three Tank System." *Proceedings of the 2nd International Conference on Circuits, Systems, Communications & Computers,* Puerto De La Cruz, Tenerife, pp. 161-166.

[19] *Digi. ZigBee RF Modules*. 2016. Accessed Apr 21, 2017. Available in: <https://www.digi.com/resources/documentation/digidocs/PDFs/90000976.pdf>.

[20] Casilari, Eduardo., Cano-García, Jose M., and Campos-Garrido Gonzalo. 2010. "Modeling of Current Consumption in 802.15.4/ZigBee Sensor Motes." *Sensors*, v. 10, pp. 5443-5468.

[21] Mansano, Raul K., Godoy, Eduardo P., and Arthur J.V. Porto. 2014. "The Benefits of Soft Sensor and Multi-Rate Control for the Implementation of Wireless Networked Control Systems." *Sensors*, v. 14, pp. 24441-24461.

In: Networked Control Systems
Editor: Eduardo Paciência Godoy
ISBN: 978-1-53613-105-5
© 2018 Nova Science Publishers, Inc.

Chapter 7

IoT-BASED NETWORKED CONTROL SYSTEMS: A PROPOSAL AND CASE STUDY

Jeferson André Bigheti[1,*], Sérgio Luiz Risso[1] and Eduardo Paciência Godoy[2]
[1]National Service of Industrial Training (SENAI), Lençóis Paulista, SP, Brazil
[2]São Paulo State University (Unesp), Sorocaba, SP, Brazil

ABSTRACT

New manufacturing applications such as Industry 4.0 have focused on the joint use of industrial automation and information technology for greater efficiency, quality and productivity. For the development and implementation of this new approach, it is necessary to apply some recent technologies such as Internet of Things (IoT), wireless network control systems (WNCS) and Cloud Computing. This work presents the proposal of an IoT-based networked control architecture based on integration of a WNCS to a Web server. Control system devices communicate via COAP protocol and the Web server is implemented using the open source tool Node.js. A case study using the Hardware-In-the-Loop (HIL) technique

* Corresponding author: Email:jefersonbigheti@terra.com.br.

evaluates the closed loop control of a level plant through a modified PID controller with communication via Web Service RESTful with the data server.

Keywords: Cloud computing, COAP protocol, Hardware-In-the-Loop, PIDPlus

INTRODUCTION

Industries are changing to improve their processes and thereby achieve greater efficiency and quality within lower costs and time. The machines already talk among themselves and with us using the concept of the Internet of Things (IoT). Expressions such a big data, decentralization, virtualization, automation and Internet of Things are concepts of automation related to the Industry 4.0. The Industry 4.0, also called the Fourth Industrial Revolution, represents an evolution of current production systems from the convergence between new technologies of industrial automation and information technology (IT) (BERGER, 2014). This concept allows greater product customization, more efficient resource utilization, lower margin of error, control of raw materials, and improved distribution logistics.

Most of the technology need for the implementation of Industry 4.0 already exist. These include the IPV6 protocol, cyber physical systems (CPS), RFID, the use of system virtualization and services from software, the Internet of Things (IoT), Cloud Computing and Big Data (ZUEHLKE, 2010; STANKOVIC, 2014, LEE ET AL., 2015, HEGAZY & HEFEEDA, 2015). The great challenge, therefore, is to promote the integration among these technologies, in order to obtain a new productive reality, where everything will be connected for the best production, cost and safety decisions, all on demand and in real time.

Solutions based on Industrial Ethernet allow the user to integrate the field equipment to IT services provided by servers (DECOTIGNIE, 2005). Use of Industrial Ethernet network protocols has expanded rapidly enabling the communication of data from factory floor and automation systems with server databases. In this way, it became possible to interconnect these automation systems through Ethernet TCP, UDP/IP with in Web applications. This interconnection of data and information network infrastructures in a high level and in a safe and reliable manner represents a trend towards the development of innovative and integrated automation and control solutions in the industrial area with Industry 4.0 (BERGER, 2014). Wireless communication is of fundamental importance for the implementation of the Industry 4.0 concept. One of the basic technologies for this concept is the Internet of Things (IoT) and the Machine to Machine (M2M) (STANKOVIC, 2014).

Cloud computing essentially refers to applications over the Internet in any place and platform as easily as having them installed on local computers (DIVYA & JEYALATHA, 2012). Kushida & Pingali (2014) presented a review of the requirements for the application of cloud computing in the industry, describing potential advantages of this technology and discussing case studies with real results of this type of implementation. Following this trend of cloud computing, there have recently been discussions about the potential of this technology for system virtualization and the creation of automation and control services in the cloud. Much research has focused on development in this direction with solutions that enable the integration of automation equipment with IT services that are stored in the cloud.

Hegazy & Heffeda (2015) presented a new proposal for a cloud automation architecture. In this proposal, several applications such as closed loop control, process monitoring and optimization would be performed through cloud services. The authors introduced a new service in the cloud applied in industrial automation where the controllers are hosted on two physically separated cloud servers. The authors state that the proposal would bring numerous benefits in terms of cost reduction and development time through virtualization of controllers, reduction of hardware used and time of application development with the use of services made available. Nikolaidis et al. (2015) proposed the interaction between two concepts: the use of controllers in the clouds with devices based on IoT. The proposal of Chen et al. (2010) introduced the use of Robot as a Service or RAAS, this being a service in the cloud to access the hardware and software of a robot. Finally,

Wu et al. (2013) showed the concept of Cloud Manufacturing, which is a production model using the definition of cloud computing.

Usually in IoT systems, the attention is concentrated on wireless sensors, cloud computing and big data techniques. However, the idea of IoT can be extended to closed loop solutions, for example with Networked Control Systems over the Internet (QUEVEDO ET AL., 2017). Even though this idea would enable a whole range of novel functionalities, feedback control design and architectures for IoT imposes significant challenges that are not addressed yet. This chapter presents an architecture proposal and an experimental case study for an IoT-based NCS with devices interconnection through an M2M server and cloud control services. The architecture is based on the application of Wireless Networked Control Systems (WNCS) using the COAP protocol and RESTful web service. Information exchange between sensors and actuators and the M2M Web server is done using the COAP protocol. The closed loop control proposed is done using the PIDPlus controller, running as a cloud service and communicating with the M2M server using RESTful technology. The architecture is tested in a case study using Hardware-In-the-Loop (HIL) with a simulated plant operating with real devices, server and communication.

The chapter is organized as followed. After this introduction, the technologies used in the paper are presented. The proposed architecture for the IoT-based control systems is detailed and explained. A case study to test the architecture is implemented using the HIL technique. The results for the case study are analyzed and discussed. At the end, the final remarks and conclusions of the paper are presented.

APPLIED TECHNOLOGIES

Internet of Things (IoT)

Kevin Ashton used the term IoT for the first time in 1999. The idea was a new technology that aimed to interconnect Procter & Gamble products through RFID transceivers to manage the company's product distribution by connecting the packaging to the Internet (ASHTON, 2009). After several years, this term is still used in many scientific articles and developments. IoT can be seen as the extension of the Internet with the interconnection sensors, actuators, machines, systems and people through the real world with the same ease and flexibility that is already done in the virtual world.

This IoT technology has gained practical importance in the last years with the technological advance (FLEISCH 2010). IoT contemplates the logical connection of all devices and means related to the productive environment, such as sensors, controllers, computers, production cells, productive planning system, and all information shared through databases. IoT devices can exchange information with each other. This information has to be stored (for example in the cloud) and to be processed, used and accessed by anything or anyone who has access to the network, regardless of their geographical location.

There are three essential elements for IoT: hardware (sensors, actuators and communication hardware), data presentation interface and middleware, which mediate between the other two elements. On the other hand, the IoT middleware consists of an on-demand storage interface and computational tools for data analysis (GUBBI et al., 2013).

M2M is an acronym for Machine-to-Machine, according to Stankovic (2014), M2M represents the interconnection between devices or systems, where such systems can autonomously exchange information between each other and make process-related production and safety decisions through an intelligence model managed by a programmable device. M2M encompasses specific devices with the ability to collect data in a given environment, which transmit the collected data to an application that is normally interconnected by servers. These devices classified as data collectors, usually sensors, may be unable to communicate directly with the application, requiring the use of another instrument, which will act as an intermediary in the communication, transmitting the data received to the other side through gateway. It is important to note that this process is executed automatically, that is, there is no previous request made by a user, which will be received by the sensor device to be processed next. Thus, it is possible to define M2M as a set of technologies, with the objective of establishing a wired or non-wired communication between sensor devices with limited capacity to transmit information to one or several applications automatically.

REST Architecture

Roy Fielding (2000), in his academic work Architectural Styles and the Design of Network-based Software Architectures at the University of California, established the principles and guidelines of what came to be known as Representational State Transfer (REST). The REST protocol is one of the main designs for Web communication, using HTTP call patterns such as GET, PUT, POST, and DELETE. The client application sends a request (HTTP Request) containing the information necessary to perform a certain operation on the server. The server processes the request and returns to the client application a response (HTTP Response), usually containing a string file in the Java Script Object Notation (JSON) protocol. The data exchanged may range from a simple message to a complex information set. The JSON protocol is a light format for data connection, represented by an object, which is formatted as follows: {text: value, text: value, ... text: value}. In this way, it is possible to easily represent a complex writing of information. In the REST architecture, there is no service descriptor. The request made by the client assumes that it knows what should be sent to the server, thus facilitating the implementation process.

COAP Protocol

A working group of the Internet Engineering Task Force (IETF) called Constrained RESTful Environments (CoRE) created the Constrained Application Protocol (CoAP). The group's activities began in March 2010 with the objective of creating a framework for applications that manipulate simple resources embedded in interconnected devices in limited networks. The CoAP is an application protocol intended to be used in electronic devices, allowing interactive communication with the Internet and other devices, with RFC 7252 being the protocol specification. One of the main goals of CoAP is to be an alternative web protocol for networks with devices with energy and memory restriction. The advantages of using an HTTP compliant protocol are the ease of integration and reuse of applications. CoAP is a REST set optimized for M2M. The IETF establishes the minimum conditions for developing an HTTP compliant application protocol, but focused on applications where power and memory are scarce. The CoAP protocol was designed taking into account the energy restrictions and high failure rates in the transmission of the WSN packets, being their asynchronous communication between the points when using the UDP.

According to RFC 7252 (IETF, 2014), the CoAP is a transfer protocol aimed at networks with small amount of RAM and in those where the rate of packet loss is high. The protocol was designed for M2M applications, for example, smart energy and industrial and residential automation. It defines four types of messages: Confirmable, Non-confirmable, Acknowledgment, and Reset (RFC 7252, 2014).

- **Confirmable** (CON), messages that need to be committed to the destination. Thus, when there is no packet loss, each message of this type results in a message of type Acknowledgment or Reset;
- **Non-confirmable** (NON), do not require confirmation of receipt. This feature is useful in the case of an application that receives constant readings from a temperature sensor in a very short space of time, where loss of one or another message is not cause for concern;
- **Acknowledgment** are messages that confirm the receipt of a Confirmable message. It is important to note that by itself, an ACK message does not indicate success or failure of any requests encapsulated in the Confirmable message;
- **Reset** indicate that another message (CON or NON) was received, but for lack of any context, it could not be properly processed. It can occur in case some device has restarted.

The CoAP protocol provides a request/response model between application endpoints, supports services and resources, which includes key Web concepts such as Uniform Resources Identifiers (URIs) and common Internet media types. In addition, the protocol is designed to interact with HTTP for Web integration. Below are some features described in RFC 7252:

- Web protocol that complies with M2M requirements in restricted environments;
- Exchange of asynchronous messages;
- URI and Content-Type support;
- Simple proxy and caching capabilities;
- HTTP mapping that allows proxies to provide access to HTTP CoAP resources evenly;
- Secure interconnection to Datagram Transport Layer Security (DTLS);
- Binding in User Datagram Protocol (UDP) with optional reliability supporting both unicast and multicast requests;
- Support for GET, POST, PUT, DELETE methods.

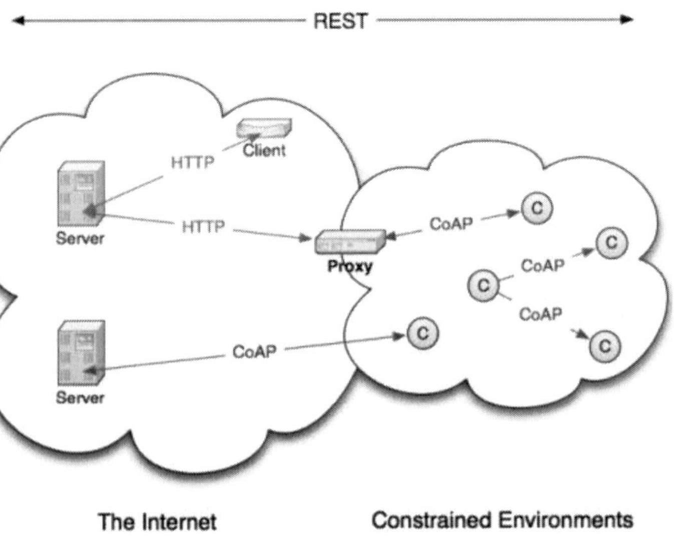

Figure 1. CoAP Architecture.

Figure 1 illustrates the CoAP architecture. One of the goals of CoRE is also to adjust the Representational State Transfer (REST) architecture to restricted environments (nodes and network). The CoAP was developed according to this architecture, so it can be considered as a RESTful protocol, according to Shelby (2014).

Figure 2 shows the format of the CoAP protocol, which is based on the exchange of messages that are transported over UDP. Messages are encoded in binary format with a 4-byte header followed by a variable-width token (0 to 8 bytes). After the token, there may be a number of CoAP options in the Type-Length-Value (TLV) format, which are optionally followed by a payload, occupying the rest of the datagram.

```
 0                   1                   2                   3
 0 1 2 3 4 5 6 7 8 9 0 1 2 3 4 5 6 7 8 9 0 1 2 3 4 5 6 7 8 9 0 1
+-+-+-+-+-+-+-+-+-+-+-+-+-+-+-+-+-+-+-+-+-+-+-+-+-+-+-+-+-+-+-+-+
|Ver| T |  TKL  |      Code     |          Message ID           |
+-+-+-+-+-+-+-+-+-+-+-+-+-+-+-+-+-+-+-+-+-+-+-+-+-+-+-+-+-+-+-+-+
|   Token (if any, TKL bytes) ...
+-+-+-+-+-+-+-+-+-+-+-+-+-+-+-+-+-+-+-+-+-+-+-+-+-+-+-+-+-+-+-+-+
|   Options (if any) ...
+-+-+-+-+-+-+-+-+-+-+-+-+-+-+-+-+-+-+-+-+-+-+-+-+-+-+-+-+-+-+-+-+
|1 1 1 1 1 1 1 1|    Payload (if any) ...
+-+-+-+-+-+-+-+-+-+-+-+-+-+-+-+-+-+-+-+-+-+-+-+-+-+-+-+-+-+-+-+-+
```

Figure 2. CoAP message format.

The fields that make up the datagram header are defined by the RFC as follows:

- **Version (ver)**: two-bit unsigned integer indicating the number corresponding to the CoAP version;
- **Type (T)**: 2-bit unsigned integer that indicates whether the message type, which may vary between Confirmable (0), Non-confirmable (1), Acknowledgment (2) or Reset (3);
- **Token Width (TKL)**: unsigned four-bit integer used to denote the token's variable width (up to 8 bytes);
- **Code (Code)** :an unsigned eight-bit integer that characterizes the message type, and may indicate a request method (Table 3) or a response code;
- **Message ID (message ID)**: 16-bit unsigned integer used to detect duplicate messages and for comparison of ACK-type messages.

The message token is used to relate to requests and responses. These messages are generated on the client and transported along with the request. The server should echo this token in the corresponding response. The header, token, and options are optional. When the message has a payload it must be indicated by a one-by-one marker (Payload Marker) indicating the end of the options and the beginning of the load. The absence of the marker means a payload of width 0 bytes, while the presence of a marker followed by a payload of 0 bytes should be processed as an error in the format of the message. Figure 3 and Figure 4 presents the codes of the methods and code of answers in the CoAP protocol.

Code	Name
0.01	GET
0.02	POST
0.03	PUT
0.04	DELETE

Figure 3. CoAP methods codes.

Code	Description
2.01	Created
2.02	Deleted
2.03	Valid
2.04	Changed
2.05	Content
4.00	Bad Request
4.01	Unauthorized
4.02	Bad Option
4.03	Forbidden
4.04	Not Found
4.05	Method Not Allowed
4.06	Not Acceptable
4.12	Precondition Failed
4.13	Request Entity Too Large
4.15	Unsupported Content-Format
5.00	Internal Server Error
5.01	Not Implemented
5.02	Bad Gateway
5.03	Service Unavailable
5.04	Gateway Timeout
5.05	Proxying Not Supported

Figure 4. CoAP response codes.

NODE.JS Server

In 2009 at a JavaScript conference, a programmer named Ryan Dahl presented a project of a platform that used a virtual machine with JavaScript V8 from Google with an event loop. This project pointed to a completely different direction from the JavaScript platforms that run on the server side: leveraging the power of JavaScript, making tasks difficult to write in asynchronous applications in easy tasks. Carter (2014) classifies Node.js as an open source platform, built on the Chrome browser's JavaScript to make it easier to build network application servers quickly and scalable applications that use a non-blocking event-driven I/O model. According to Bosák et al., (2015), the characteristics of Node.js makes it lightweight and efficient, ideal for real-time applications with intense data exchange across distributed devices. Figure 5 shows the operating principle of the NODE JS event loop.

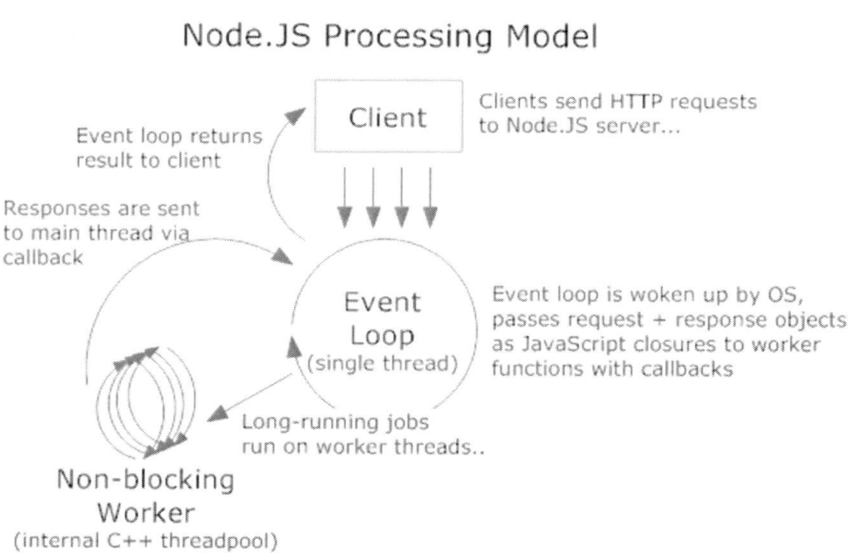

Figure 5. Functional principle of the NODE.JS event loop.

PIDPLUS Controller

The introduction of wireless sensors in industry gives rise to new challenges when variable measurement is used in control applications. To address these challenges, the PID controller can be modified to operate properly with large periods of sensor sampling, non-periodic variable updates, and loss of transmitted messages (BLEVINS et al., 2014). The key to understanding how the PID should be modified is to perform the reset of the implemented PID using a feedback network and a filter where the time constant is a direct reflection of the dynamic response of the process.

The PIDPlus implementation (SONG et al., 2006) for WNCS is illustrated in Figure 6. The PIDPlus maintains the control signal at the last calculated level until a new measure is received. It is important to note that its tuning is independent of the sampling period, depends only on the physical characteristics of the plant.

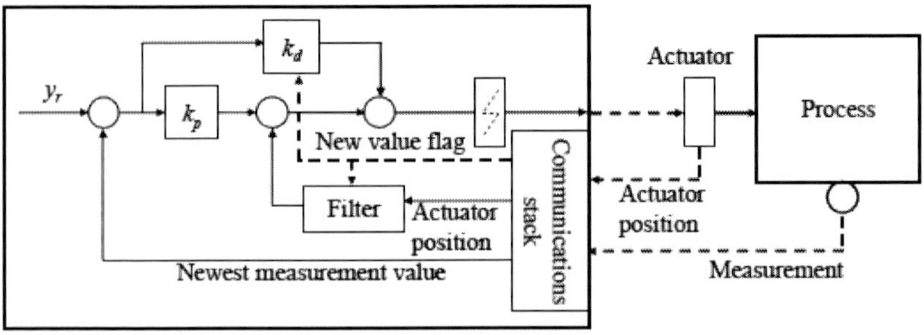

Figure 6. PIDPlus controller structure (SONG et al., 2006).

The feedback (*newest measurement value*) and the filter of the first order (*filter*) are modified to create the replenishment contribution with the following behavior:

- Keep the last output of the calculated filter (FN-1) until a new measurement is entered (*new value flag*);
- When a new measurement is received (*new value flag*), use the new filter output as feedback contribution (*FN*).

The main difference of the PID and PID PLUS is in the integrative part that has been replaced by a 1st-order filter. The filter output is calculated as (1):

$$F_N = F_{N-1} + (O_{N-1} - F_{N-1})\left(1 - e^{\frac{-\Delta T}{T_{reset}}}\right) \tag{1}$$

At where: F_N = new filter output, F_{N-1}= filter output at last run, O_{N-1}= controller output on last run and ΔT = time interval since the last measured value has been received and T_{reset} = plant time constant added to the dead time.

The derivative part is replaced as (2):

$$O_D = K_D \frac{e_N - e_{N-1}}{\Delta T} \tag{2}$$

At where: e_N= current error, e_{N-1}= last mistake, O_D = derivative of the controller and K_D = derivative gain.

Consider the contribution of the derivative when inputs are lost for several periods. For the traditional PID algorithm, the divisor in the derivative part would be the period (discretization of the controller), while in the PIDPlus algorithm the time elapsed between two measurements received successfully (ΔT). It is obvious that the modified algorithm produces a derivative action smaller than the PID control algorithm. (SONG et al., 2006).

In the implementation of PIDPlus, the reset calculation automatically compensates for the change in measurement and refresh rate of the measurement. Derivative term calculations for a new measurement value are not available at each PID run. Thus, there is no need to modify the synchronization for the wireless control, i.e., the adjustment is based strictly on the gain and dynamics of the process.

Hardware-In-the-Loop

Studying characteristics of systems through simulation has gained importance over the years, and this practice ensures significant benefits to automation and control companies, such as reducing costs with prototyping, possibility to testing systems under different conditions with high repeatability; and as a result optimizing processes and product development (Sahin et al., 2012). There are different techniques and purposes for conducting process control simulations. The three most known simulation techniques are the Rapid Control Prototyping (RCP), the Hardware-In-the-Loop (HIL) and the Software-In-the-Loop (SIL) (ISERMANN, 2008). RCP and HIL are considered real-time simulations, as there is a requisite for a real-time operational system and dedicate hardware for models to provide deterministic temporal behavior.

Hardware-in-the-loop (HIL) refers to a system where some of the components are real rather than simulated. Considering the HIL technique, usually real equipment (controller, communication) are connected to a virtual plant executed on a real-time simulator. The main advantages of HIL is allowing reliable tests in conditions unavailable on real plants. Mathematical models can replace real components and the components to be tested are inserted into the closed loop.

The HIL simulation technology can be found in almost all areas where more realistic testing of system components is required before its final construction. HIL simulation is also a very useful tool for evaluating and developing controllers, offering a zero risk in the experimentation of different techniques and control methodologies without the need of real platform for testing. In this way, it is possible to save investments and avoid dangerous consequences resulting in errors in the design phase.

The context of HIL simulation shows advantages of using real-time models emulation in a highly controlled environment such as the IoT and NCS with low cost implementation. Another objective of this paper is to show how to use HIL simulations for an IoT-based networked NCS for design and tests.

PROPOSED ARCHITECTURE

The IoT-based control architecture proposed in this paper is shown in Figure 7. A simulated process of a level plant runs in the LabVIEW Realtime using the Hardware-In-the-Loop technique. The process model used a National Instruments MyRIO hardware with realtime capabilities. The industrial cloud (Server block) uses Node.js as a Web Server, storing the control data used in the process in a mySQL database. The Controller block is responsible for controlling the process by accessing the control variables using a Web Service communication of Node.js. The control algorithm used is the PIDPlus presented. The PIDPlus is a modified version of the traditional PID controller for wireless control applications. This web service uses REST technology in a client-server architecture accessible via the TCP/IP network. In addition, it would also be possible to implement the controller as a web server service (e.g., via Java script) instead of developing it in the LabVIEW environment. This architecture also enables and standardizes the web service communication between the server and different applications that would be developed such as supervision and optimization.

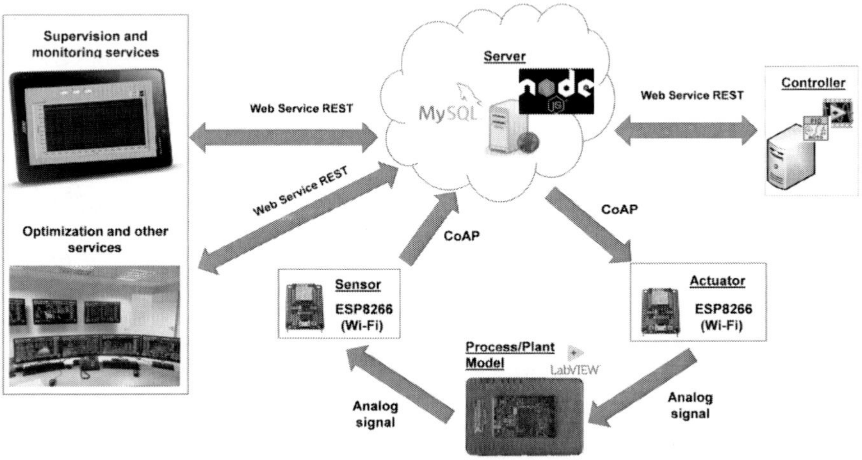

Figure 7. Architecture for Cloud Automation and Control.

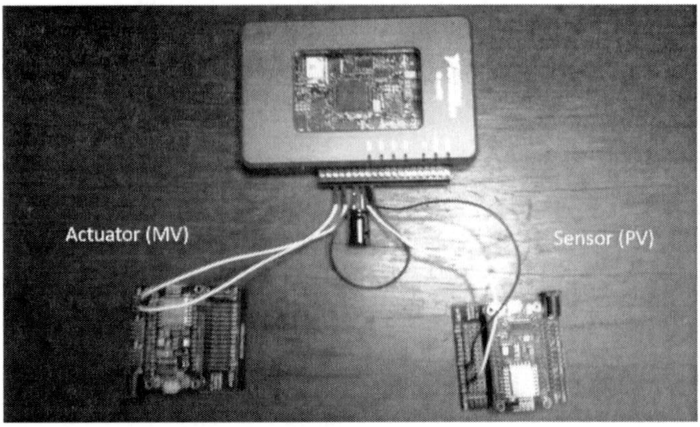

Figure 8. HIL connection in the proposed architecture.

The Sensor and Actuator blocks of the architecture are composed by NodeMCU hardware with the ESP8266 module. The NodeMCU implements the CoAP protocol required for the communication with the Server block. The Sensor block acquires (ADC) the process variable (PV) using an analog signal from a DAC output of the MyRIO. The Actuator plant is responsible to receive the control signals calculated by the Controller block and made available at the Server. The Actuator also uses an analog signal to act on the plant (MyRIO). A detailed view of the connection among the process, sensor and actuator is presented in Figure 8.

The control signal (MV) in the Server block, upgraded by the Controller block, is transmitted to the Actuator block via COAP protocol using a Wi-Fi connection. The actuator block uses an ESP8266 Wi-Fi card. The proposed system enables the IoT for automation and process control by providing online, sensor feedback information and control signal to actuators for any type of platform, such as smartphones, tablets or computers, making monitoring and control data accessible in real time, for any point connected to the network and/or the Internet.

The prosed architecture provides great flexibility and can be expanded to other types of processes models, control algorithms and simultaneous control loops, predicting the global real behavior of the IoT systems, reducing risks, cost and time with tuning and startup of automation projects.

EXPERIMENTAL SETUP AND RESULTS

In order to evaluate the proposed architecture for IoT-based NCS, a case study is done. The case study is based on a simulated level control process using the HIL scheme presented in the architecture of Figure 11. The proposed process of a level control system is presented in Figure 9. The problem was to develop the PIDPlus controller to control the level h of the tank by manipulating the inlet valve its flow q_{in}.

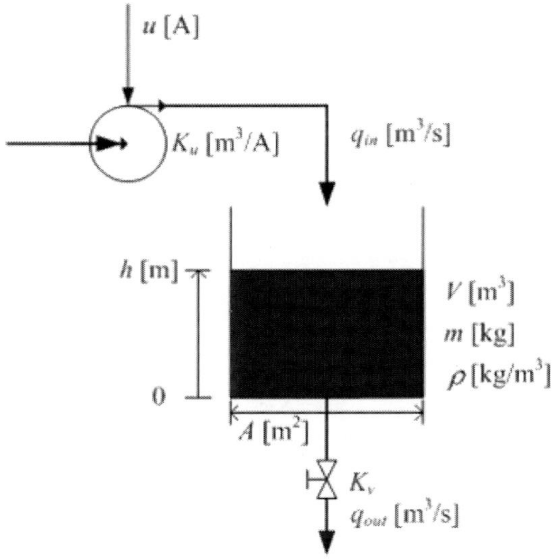

Figure 9. Level Control Process for the Case Study.

where: $q_{in}(t)$ is the input flow, $q_{out}(t)$ is the output flow, ρ is the density = 1000, m is the mass, V is the volume, A is the cross section area of the tank = 1, $h(t)$ is the height of te liquid in the tank, g is the gravity, K_u is the pump gain = 5, u is the control signal of the pump, and K_v is the valve constant = 0.005.

The mathematical model of the level process given in Figure 10 was implemented with the LabVIEW Control Design and Simulation and runs in real-time in the MyRIO hardware.

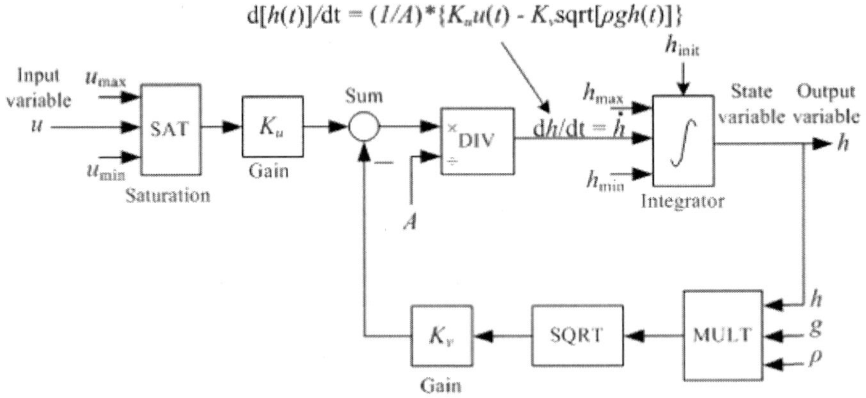

Figure 10. Level Control Mathematical Model for the Case Study.

The differential of the IoT control system is that the controller is physically separated from the process, obtaining the closed loop data from the Server in Node.js. Using the web service via REST, the cloud data is used and the controller is able to control the process. The REST web service allows extending the Server functionalities through the client/server architecture, where Node.js acts as server, and an external module (Controller block) acts as client. The Node.js and the Client application can run on the same machine, or on separate computers, as long as they are connected via the IP network.

A Wi-Fi network between the sensor and an Access Point was created. The Node.js server (process data) with the mySQL database stores the process control data. For the case study, the Server provides the following variables:

- Hardware-In-the-Loop Module – PV;
- Hardware-In-the-Loop Module – MV;
- Hardware-In-the-Loop Module – Setpoint;
- Hardware-In-the-Loop Module – P (Proportional);
- Hardware-In-the-Loop Module - I (Integral);
- Hardware-In-the-Loop Module - D (Derivative).

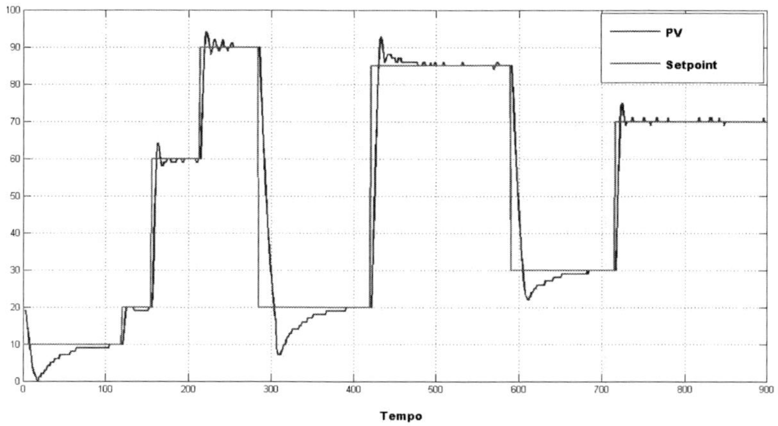

Figure 11. Results of the PIDPlus Controller for the HIL level control using the IoT architecture.

The analog MV control signal (0 to 3.3Vdc) generated from a PWM signal of the ESP8266 module controls the actuator of the HIL input. The Controller block accesses the sensor data (PV) generated by the HIL module on the cloud web server and implements the control algorithm. The calculated control signal is updated on the server. All communication between Controller and Web server is done via Web Service REST. The controller used was the PIDPlus (BLEVINS et al., 2014). The closed-loop control cycle implemented in the proposed architecture has an update rate of 2s. The CoAP (UDP over Wi-Fi) was used for communication between the sensor/actuator and the Web server.

The behavior of the PIDPlus controller for level control through the cloud server can be seen in Figure 11. The initial results validate the implementation the IoT –based NCS. Even though more experiments have be done with the developed architecture, Figure 11 proves that a NCS can be applied to control a process using COAP and REST communication. Future work aims to implement the designed controller in the process control pilot plant in order to validate the results obtained with the HIL technique.

CONCLUSION

In this paper, a WNCS communicating via COAP and REST was integrated with a Web server in a closed loop control system. The resulted architecture for IoT-based NCS was implemented and evaluated using a HIL technique. This new type architecture in which control systems are integrated with IoT infrastructures represents the next evolution of networked control architectures.

The use of HIL technique for the design and test of these IoT-based NCS provided flexibility and allowed studies to be conducted, for example the design and implementation of control strategies.

The use of the PIDPlus controller proved to be effective for the IoT case study because it incorporates modifications in the traditional PID structure to compensate for the impact of non-periodic samples and time-varying delays originated through the insertion of the web server in the closed loop, and to compensate for possible loss of data transmitted. The performance of the level control was adequate considering the characteristics of the process.

The use of the NODE.js as a web server complies with the requirements of the application, as well as providing the possibility of using other tool features in the proposed architecture such as supervision, alarm management and reporting. The communication via Web Service REST implemented between the controller and the web server was functional and simple, showing potential of use for other purposes. The communication via CoAP between the sensor/actuator and the web server was simple, reducing the necessary bandwidth for data transmission according to the HTML standard.

The integration of automation and IT technologies such as the IoT in a networked control architecture presented in this paper may allow the analysis and decision-making based on the data available in the cloud, beyond the monitoring and improvement of process efficiency. This paper presented preliminary results, but allows us to glimpse potential applications of the developed architecture, which will be studied in future works, such as:

- Cloud control applications;
- Providing system identification and advanced control services using historical data available on the process web server;
- Monitoring and optimization of control processes using cloud services;
- Virtualization and redundancy of controllers;
- Possibility of cost reduction and hardware used, in addition to time of development of applications with the use of services made available as virtual commissioning services.

The concepts and results of this paper can be expanded to other types of processes models, control algorithms and simultaneous control loops, predicting the global real behavior of the IoT systems, reducing risks, cost and time with tuning and startup of automation projects.

REFERENCES

Ashton, Kevin. (2009). "That Internet of Things Thing: In the real world, things matter more than ideas." *RFID Journal.* Accessed September 3, 2016. http://www.rfidjournal.com/articles/view?4986.

Berger, Roland. (2014). *"Industry 4.0 – The new industrial revolution, Strategy Consultants."* Accessed November 2015. https://www.rolandberger.com/media/pdf/Roland_Berger_TAB_Industry_4_0_20140403.pdf.

Blevins, Terry., Mark, Nixon. & Willy, Wojsznis. (2014). *"PID Control Using Wireless Measurements"* Paper presented at the American Control Conference (ACC). Portland, Oregon, June 4-6.

Bosák, Tomáš. & Katarína, Žáková. (2015). *"Node.js based remote control of thermo-optical plant."* Paper presented at the 12th International Conference on Remote Engineering and Virtual Instrumentation (REV), Bangkok, Thailand, February 25-27. IEEE.

Carter, Bryan. (2014). *"HTML Educational Node.js System (HENS): An Applied System for Web Development."* Paper presented at the annual

Global Online Conference on Information and Computer Technology, Louisville, Kentucky, December 3-5. IEEE

Decotignie, Jean-Dominique. (2005). "Ethernet-based real-time and industrial communications." *Proceedings of the IEEE*, *93*, 1102-17.

European Telecommunications Standards Institute. (2013). *"TS 102690 version 2.1.1: Technical Specification"*. Accessed August 15, 2017. <http://www.etsi.org/deliver/etsi_ts/102600_102699102690/02.01.01_/60/ts_102690v020101p.pdf>.

Fielding, Roy T. (2000). *"Architectural Styles and the Design of Network-based Software Architectures."* PhD diss., University of California.

Fleisch, Elgar. (2010). *"What is the Internet of Things? An Economic Perspective"*, Monograph, University of St.Gallen.

Gubbi, Jayavardhana., Rajkumar, Buyya., Slaven, Marusic. & Marimuthu, Palaniwami. (2013). "Internet of Things (IoT): A vision, architectural elements, and future directions." *Future Generation Computer Systems*, *29*, 1645-60.

Hegazy, Tamir. & Mohamed, Hefeeda. (2015). "Industrial Automation as a Cloud Service." *IEEE Transactions on Parallel and Distributed Systems*, *26*, 2750-63.

IEEE Working Group for WLAN Standards. (1999*). "IEEE standard for wireless LAN medium access control (MAC) and physical layer (PHY)."* Accessed October 6, 2017. http://grouper.ieee.org/groups/ 802/11.

IEEE Working Group for WLAN Standards. (2003). "IEEE 802.11g: Further Higher Data Rate Extension in the 2.4 GHz Band." *IEEE.*

IEEE Working Group on Broadband Wireless Access Standards. (2004). *"IEEE 802.16 WirelessMAN standard for wireless metropolitan area networks."* Accessed October 6, 2017. http:// grouper.ieee.org/groups/802/16.

Internet Engineering Task Force. (2010). *"Constrained RESTful environments: charter core."* Accessed August 10, 2017. http://datatracker.ietf.org/doc/charter-ietf-core/.

Internet Engineering Task Force. (2014). *"Constrained application protocol: request for comments 7252."* Accessed August 10, 2017. https://tools.ietf.org/html/rfc7252.

Isermann, R. Mechatronic systems—Innovative products with embedded control, *Control Engineering Practice*, Vol. *16*, No. 1, pp. 14-29, 2008.

Kushida, Takayuki. & Gopal, S. Pingali. (2014). "Industry Cloud - Effective Adoption of Cloud Computing for Industry Solution." *Paper presented at the IEEE 7th International Conference on Cloud Computing*, Anchorage, Alaska, June 27-July 2.

Lee, Jay., Behrad, Bagheri. & Hung-An, Kao. (2015). "A Cyber Physical Systems architecture for Industry 4.0 based manufacturing systems." *Manufacturing Letters*, *3*, 18-23.

Quevedo, Daniel., Findeisen, Rolf., Ishii, Hideaki., Johansson, Karl. H. & Samad, Tariq. (2017). "Special Issue on Feedback Control for the Internet of Things". *IEEE Internet of Things Journal*, Available in: <http://iot-journal.weebly.com/uploads/1/8/8/0/ 18809834/ ieee_iot_ journal_si_feedback_control_cfp.pdf>.

Şahin, S., İşler, Y. & Guzelis, C. A Real Time Simulation Platform for Controller Design, Test and Redesign. In: *Real-Time Simulation Technologies: Principles, Methodologies, and Applications*, Chapter 19, ed. Katalin Popo-vici, Pieter J. Mosterman., pp. 461–500, CRC Press, 2012.

Shelby, Zach. (2014). "*CoAP: The Web of things protocol.*" Accessed August 10, 2017. http://pt.slideshare.net/zdshelby/coap-tutorial.

Song, Jianping., Aloysius, K. Mok., Deji, Chen., Mark, Nixon., Terry, Blevins. & Willy, Wojsznis. (2006). "Improving PID control with unreliable communications." *Paper presented at the ISA EXPO Technical Conference.* Houston, Texas, October 17-19.

Stankovic, John A. (2014). "Research Directions for the Internet of Things." *Internet of Things Journal*, *1*, 3-9.

Zuehlke, Detlef. (2010). "Smart Factory—Towards a factory-of-things." *Annual Reviews in Control*, *34*, 129-38.

About the Editor

Eduardo Paciência Godoy, PhD
Assistant Professor
Control and Automation Engineering
São Paulo State University (Unesp),
Sorocaba, SP, Brazil
Email: epgodoy@sorocaba.unesp.br
Phone: +55 15 3238 3469

Eduardo Paciência Godoy holds a Bachelors degree in Control and Automation Engineering at Federal University of Itajubá - UNIFEI (MG-Brazil) in 2003, Masters degree (2007) and PhD (2011) in Mechanical Engineering at University of São Paulo at São Carlos EESC-USP (SP-Brazil). He was a visiting researcher (2010, 2014) at the Engineering Research Center (ERC) for Reconfigurable Manufacturing Systems (RMS) at the University of Michigan (EUA), and has been a Professor at the Control and Automation Engineering course at the UNESP - São Paulo State University since 2011. He is a member of the IEEE (Institute of Electrical and Electronics Engineers) and of the SBA (Brazilian Society of Automatic). His experience and interests include manufacturing automation, industrial networks and informatics, networked control systems (NCS), wireless networks and telemetry, embedded electronics, Internet of Things (IoT) and Industry 4.0.

INDEX

O

P

R

S

T

V

W

X